FABERGÉ REVEALED
THE COLLECTION OF THE VIRGINIA
MUSEUM OF FINE ARTS

玲珑万象
美国弗吉尼亚美术馆藏法贝热珍品集

故宫博物院 编
COMPILED BY THE PALACE MUSEUM

故宫出版社
THE FORBIDDEN CITY PUBLISHING HOUSE

图书在版编目（CIP）数据

玲珑万象：美国弗吉尼亚美术馆藏法贝热珍品集/故宫博物院编. —北京:故宫出版社，2016.4
ISBN 978-7-5134-0848-6

Ⅰ．①玲… Ⅱ．①故… Ⅲ．①宝石-工艺美术-俄罗斯-现代-图集 Ⅳ．①TS934.3-64

中国版本图书馆CIP数据核字（2016）第057190号

玲珑万象——美国弗吉尼亚美术馆藏法贝热珍品集
故宫博物院编

出 版 人：王亚民
责任编辑：方　妍　李园明
装帧设计：王　梓　于朝娟
出版发行：故宫出版社
　　　　　地址：北京东城区景山前街4号　邮编：100009
　　　　　电话：010-85007808　010-85007816　传真：010-65129479
　　　　　网址：www.culturefc.cn
　　　　　邮箱：ggcb@culturefc.cn
制版印刷：北京雅昌艺术印刷有限公司
开　　本：889毫米×1094毫米　1/16
印　　张：20.5
版　　次：2016年4月第1版
　　　　　2016年4月第1次印刷
印　　数：1-1500册
书　　号：978-7-5134-0848-6
定　　价：420.00元

目 录

故宫博物院院长祝辞 单霁翔 007
美国弗吉尼亚美术馆馆长祝辞 亚历山大·纳哲斯 009

前　言 015

图版目录 017
图　版 027
沙皇的礼物 029
美轮美奂的装饰艺术 073
日常的奢华 169
信仰的力量 267

专　论 297
伪法贝热　格扎·冯·哈布斯堡 298
巧夺天工，至精至美——法贝热装饰艺术　宋海洋 316

后　记 327

Contents

Foreword by the Director of the Palace Museum   Shan Jixiang 008

Foreword by the Director of the Virginia Museum of Fine Arts   Alex Nyerges 012

Introduction 016

List of Plates 022

Plates 027

Gifts of Tsars 029

Majestic Decorative Arts 073

Everyday Opulence 169

Power of Faith 267

Essays 297

Fauxbergé   Geza Von Habsburg 306

Superb Craftsmanship and Amazing Exquisiteness:

The Decorative Art of Fabergé   Song Haiyang 321

Postscript 328

故宫博物院院长祝辞

欧洲有许多历史古老且充满传奇的珠宝品牌，"法贝热"是精湛工艺、奢侈珠宝以及俄罗斯皇室珍宝的代名词。俄罗斯著名金匠、珠宝首饰匠人、工艺美术设计家彼得·卡尔·法贝热（1846～1920年）被沙皇钦点为御用金匠与珠宝师，由他设计制作的"复活节彩蛋"不仅是俄罗斯珠宝史上，也是世界珠宝史上的惊世奇作，被沙皇俄国、各国皇室和世界诸多博物馆视为珍品而珍藏。

弗吉尼亚美术馆成立于1936年，现有藏品超过33000件，藏品历史年限跨越5000年，涵盖了世界上诸多重要文明，是公认的美国顶尖综合性艺术博物馆之一。在弗吉尼亚美术馆的永久馆藏中，包括除俄罗斯以外数量最多的彼得·卡尔·法贝热的作品，以及全美最优秀的艺术品收藏之一、英国著名的银器、印象派作品、后期印象派作品、英国体育和当代艺术作品，还有著名的东亚、南亚、喜马拉雅地区和非洲的艺术品。

2011年，故宫博物院与美国弗吉尼亚美术馆在北京签署了两馆合作框架意向书，使双方的交流与沟通合作迈入了新的阶段。2014年10月18日至2015年1月19日，两馆在弗吉尼亚美术馆联合举办了"紫禁城——北京故宫博物院皇家珍品展"，通过题材广泛的精美展品，带领美国观众走进明清帝后们的生活，近距离体验博大精深的中国传统文化，更深入地了解中国传统文化的精髓和内涵。

今天，作为最为重要的中美文化交流的盛事，"玲珑万象——来自美国的俄罗斯皇家法贝热装饰艺术展"，为观众呈现234件（套）法贝热珍品，这是美国博物馆的藏品首次大规模在故宫博物院展出。展览通过沙皇的礼物、美轮美奂的装饰艺术、日常的奢华、信仰的力量四个单元系统展示了法贝热珠宝卓越的理念与工艺，为观众献上别致精美的文化体验。

文化风格千秋，艺术恒久无界。多元化的今天，古老的紫禁城敞开胸怀，迎接着五湖四海的朋友。传统文化的精髓与奔腾汹涌的时代浪潮在碰撞中交流，在交流中传播，汇聚成这个时代的文化强音。

感谢弗吉尼亚美术馆、故宫博物院双方工作人员为展览的成功举办所付出的辛苦努力。在此，我谨代表故宫博物院，并以我个人的名义祝展览取得圆满成功。

Foreword by the Director of the Palace Museum

Shan Jixiang

Europe has many time-honored and legendary jewelry brands. "Fabergé" is synonymous with exquisite craftsmanship, luxury jewelry and Russian imperial treasure. Peter Karl Fabergé (1846-1920) was a noted Russian goldsmith, jeweler, and designer of arts and crafts. He was appointed by the Tsar himself as the royal goldsmith and jewelry designer, and the Fabergé eggs he designed and made as Easter gifts were masterpieces in the Russian as well as world jewelry history. They were seen as treasure by imperial families of the Russian Empire and other countries, and collected by museums all over the world.

Established in 1936, the Virginia Museum of Fine Arts (VMFA), recognized as one of the top comprehensive art museums in the United States, has a collection size of more than 33,000 works that span 5,000 years and cover a number of important civilizations in the world. Permanent collections at VMFA include the largest Fabergé collection outside of Russia, one of the best art collections in the U.S., famous British silverware, impressionist and post-impressionist works, works about British sports and modern art, and well-known artworks from East Asia, South Asia, the Himalaya region and Africa.

In 2011, the Palace Museum and VMFA signed a Memorandum of Understanding in Beijing, bringing the exchange, communication and cooperation between the two museums to a new level. From October 18, 2014 to January 19, 2015, the two museums co-hosted the "*Forbidden City: Imperial Treasures from the Palace Museum, Beijing*" in the VMFA. The exquisite exhibits featuring a wide range of themes took the American visitors into the life of the emperors and empresses of the Ming and Qing dynasties, allowed them to feel at close range the profound traditional Chinese culture and understand its essence and connotations more deeply.

Today, as the most important event for China-U.S. cultural exchange, the Palace Museum and VMFA present to the audience the "*Fabergé Revealed*", at which 234 pieces (sets) of Fabergé treasure will be exhibited, the first time a large collection from a U.S. museum is seen in the Palace Museum. The exhibition, which comprises four sections – "Tsars' Gifts", "Majestic Decorative Arts", "Everyday Opulence" and "Power of Faith", is a systematic display of the excellent philosophy and craftsmanship of Fabergé jewelry, offering the audience a fabulous cultural experience.

There is no limit to the variety of cultures, just as there is no limit or boundary to art. In today's world of diversity, the ancient Forbidden City is opening its arms to welcome friends from far and wide. The essence of traditional culture and the billowing waves of the time merge amid collisions and spread through exchange, giving rise to the strongest note of this age.

I would like to express my gratitude to staffs of both VMFA and the Palace Museum for their hard work to ensure the success of the exhibition. On behalf of the Palace Museum and myself, I wish the exhibition a great success.

美国弗吉尼亚美术馆馆长祝辞

亚历山大·纳哲斯

弗吉尼亚美术馆的藏品浩如烟海、灿若明星。杰出的馆藏包含逾33000件艺术珍品,几乎涵盖了世界上所有重要文明。最引人瞩目的藏品包括:悉尼·刘易斯和弗朗西斯·刘易斯夫妇捐赠的新艺术、装饰派艺术和现当代美国艺术藏品;保罗·梅隆夫妇捐赠的法国印象派和后印象派艺术藏品,以及英国体育运动艺术藏品;通过哈伍德和路易丝·科克伦基金收集的美国艺术珍藏品;杰罗姆·甘斯和丽塔·甘斯夫妇收藏的英国银器;以及莉莲·托马斯·普拉特收藏的法贝热珠宝。弗吉尼亚美术馆收藏的南亚、喜马拉雅地区和非洲艺术品是全美最好的收藏之一。本馆的东亚藏品主要包括中国、日本和韩国的绘画、版画、雕塑、陶瓷、金属制品、漆器和其他装饰艺术,代表着纵贯4500年的视觉历史,收藏主题包括青铜器时代、佛教的传播、跨文化影响、陶瓷发展与贸易、文房四宝、武士文化,以及茶道和香道艺术。这些作品揭示了东亚各国之间及其与其他国家之间丰富的文化影响和交流互动。

19世纪末和20世纪初,彼得·卡尔·法贝热和他的公司专为俄罗斯帝国的王公贵族制作精美的饰物,这些饰物被认为是有史以来最为精致的珠宝艺术品。时至今日,法贝热的珐琅相框、时钟、香烟盒、手杖柄、水晶动物雕塑、玉石花卉、镶满珠宝的胸针和首饰盒,仍像当年在圣彼得堡、莫斯科和伦敦的法贝热展示厅橱窗中展出时一样,令观者赞不绝口。其中最具魅力的仍属美轮美奂的俄罗斯复活节彩蛋。这些由俄罗斯皇室特别委托法贝热制作的复活节彩蛋,因其独创性和纯观赏性,而被誉为旷世无双的珍品。

弗吉尼亚美术馆的法贝热藏品、俄罗斯装饰艺术和皇家复活节彩蛋珍藏是俄罗斯境外规模最大的公共收藏。这批藏品大都是弗吉尼亚州弗雷德里克斯堡的莉莲·托马斯·普拉特(1876~1947年)于1933年至1946年间收集而来的。普拉特夫人是通用汽车公司一位高层管理人员的妻子,对罗曼诺夫王朝的历史和法贝热艺术品怀有极大的热情,但这些作品并不一定迎合她丈夫的艺术品位。她对收藏的果敢超越了与其同时代的富有的美国女性收藏家。为了买下彼得大帝彩蛋，她竟然支付了33期月供。莉莲的这笔收藏在1947年遗赠给了一家新成立的机构，即1936年向公众开放的弗吉尼亚美术馆。

弗吉尼亚美术馆的法贝热收藏前往故宫博物院的旅程，开始于2011年。当时时任弗吉尼亚州州长的罗伯特·麦克唐纳阁下，同我一道在北京，与时任故宫博物院院长郑欣淼签署了谅解备忘录，商定了展览交流和博物馆人员交流事宜，这标志着两馆之间长期、多层面的合作正式启动。

这一历史性的协议被看做是弗吉尼亚州与中华人民共和国之间文化和商业纽带的一部分，并获得了现任弗吉尼亚州州长特里·麦考利阁下和现任故宫博物院院长单霁翔的大力支持。它的签署与故宫博物院前常务副院长李季的远见和努力是分不开的。自2010年对我馆进行访问以来，李院长对此始终予以大力支持。这些交流已经取得显著成果，对于有幸与海外同行进行交流与分享的双方博物馆工作人员而言更是如此。2014年10月18日至2015年1月19日，具有里程碑意义的"紫禁城——北京故宫博物院皇家珍品展"在弗吉尼亚美术馆展出，为这一展览交流项目拉开了序幕。娄玮副院长前来我馆参加了开幕式，同时于2015年4月出席了在华盛顿举办的"中美文化论坛"，并就两馆间的成功合作做了介绍。今天，"玲珑万象——来自美国的俄罗斯皇家法贝热装饰艺术展"即将开幕，这将是美国博物馆的藏品首次在故宫博物院展出。这一精彩纷呈、富于启迪的合作项目正值一个特殊的历史时刻，即故宫博物院成立90周年，法贝热藏品将在新近修缮完成的午门展厅展出。

在中华人民共和国驻美大使崔天凯以及华盛顿特区使馆工作人员，特别是前驻美国使馆公使陆慷和驻美国使馆文化处公使衔参赞李鸿的努力下，此项目最终得以成行。弗吉尼亚美术馆的各界友人也充分认识到增强中美文化纽带的重要意义，对此项目给予了慷慨的帮助。

我们要感谢故宫博物院的工作人员，特别是外事处李绍毅、袁宏、王蕾和王丝滢；正是在他们的不懈努力下，莉莲·托马斯·普拉特收藏的200多件精美的法贝热艺术品和俄罗斯装饰艺术品才得以来到北京。我们还要感谢故宫出版社的工作人员为出版展览图录所做出的贡献；感谢故宫博物院展览部的工作人员，包括展览部的纪炜副主任、宋海洋副研究馆员、王博馆员和李怀玉馆员为策划本次展览所付出的努力。

弗吉尼亚美术馆方面，负责1890年至今装饰艺术的主管巴瑞·希夫曼先生负责管理莉莲·托马斯·普拉特的收藏，他在客座策展人、法贝热专家格扎·冯·哈布斯堡的协助下负责该项目的筹备。首席主管兼负责艺术与教育的副馆长迈克尔·泰勒博士，领导着本馆的筹展团队。此外，负责设施与藏品管理的副馆长斯蒂芬·博纳迪斯先生负责管理随展团组。自2012年起，法贝热布展团队成员已随展访问了其他五个巡展场地，他们包括：馆藏总管南希·尼科尔斯、艺术品操作部门负责人杰夫·斯特朗、器物修复室主管希拉·帕亚奎，以及文物修复技师珍·布里奇斯。展览协调员吉娜·卡瓦罗·柯林斯负责展览的主要组织工作。首席财务官侯赛因·萨迪德负责此项目的财务管理工作。最后，还要感谢我们东亚艺术研究员李建，感谢她对此次合作的大力推动，对各项人员交流的不懈支持，以及对"紫禁城——北京故宫博物院藏皇家珍品展"的成功策划。

最后，我要感谢弗吉尼亚美术馆和故宫博物院的全体员工及支持者在本图录制作过程中的出色工作。在此，我谨代表每一位参与该合作项目及以往交流活动的人士，邀您共赏这本精美的纪念册；无论是对于拥有80年历史的弗吉尼亚美术馆还是拥有90年历史的故宫博物院来说，此次展览都具有极其重要的意义。

Foreword by the Director of the Virginia Museum of Fine Arts

Alex Nyerges

The Virginia Museum of Fine Arts houses a remarkable permanent collection of more than 33,000 works of art from almost every major world culture. Especially noteworthy are the museum's collections of Art Nouveau, Art Deco, and Modern and Contemporary American art donated by Sydney and Frances Lewis; French Impressionist and Post-Impressionist art and British sporting art given by Mr. and Mrs. Paul Mellon; American art acquired through the J. Harwood and Louise B. Cochrane Fund; The Jerome and Rita Gans Collection of English Silver; and The Lillian Thomas Pratt Collection of Fabergé jeweled objects. The museum's holdings of South Asian, Himalayan, and African art are among the finest in the United States. Representing 4,500 years of visual history, the VMFA's East Asian collection features paintings, prints, sculptures, ceramics, metalwork, lacquers and other decorative arts from China, Japan and Korea. Collection themes include the Bronze Age, the spread of Buddhism, cross-cultural influences, ceramic development and trade, scholars' implements, samurai culture, as well as tea-ceremony and incense art. These works reveal rich cultural influences and interactions both within East Asia and beyond.

The exquisite objects created by Peter Karl Fabergé and his firm in the late nineteenth and early twentieth centuries for the aristocracy and nobility of Imperial Russia are considered to be among the most refined jeweler's art of any age. The array of enameled picture frames, clocks, cigarette cases and cane handles, rock-crystal animals, hardstone flowers, and jewel-encrusted brooches and boxes still evoke the same fascination that they did when displayed in the windows of Fabergé's showrooms in St. Petersburg, Moscow, and London. Of greatest fascination are the extraordinary Easter eggs created as special commissions for the Russian imperial family works that remain unparalleled in their ingenuity and sheer beauty.

The Virginia Museum of Fine Arts' collection of Fabergé and Russian decorative arts and Imperial Easter eggs is the largest public assemblage of outside of Russia. The collection was largely formed between 1933 and 1946 by Lillian Thomas Pratt (1876-1947) of Fredericksburg, Virginia. The wife of a General Motors executive, Mrs. Pratt was passionate about the history of the Romanovs and about Fabergé objects, which were not necessarily to the taste of her husband. Her sheer determination prevailed over the lavish budgets of her American female rivals: she paid for the Peter the Great Egg in 33 monthly installments. Her collection was bequeathed to a new institution in 1947, the Virginia Museum of Fine Arts, which had opened its doors in 1936.

The Virginia Museum of Fine Arts' Fabergé Collection began its journey to the Palace Museum through a multiyear, multifaceted collaboration formally inaugurated in 2011 when the Honorable Robert F. McDonnell, then governor of Virginia, joined me in Beijing with Zheng Xinmiao, director of the Palace Museum, to sign a Memorandum of Understanding that encompassed an exchange of exhibitions and museum staff.

Envisaged as part of a greater cultural and commercial relationship between the Commonwealth of Virginia and the People's Republic of China—and warmly endorsed as such by our new Virginia governor, the Honorable Terry McAuliffe, and the new director of the Palace Museum, Shan Jixiang—this historic agreement was made possible by the vision and efforts of Li Ji, former executive deputy director of the Palace Museum who supported this endeavor from its inception during a 2010 planning visit. The benefits of the exchange have been significant, especially for the staff of the two museums that have had the opportunity to share information with their counterparts. The landmark exhibition, *Forbidden City: Imperial Treasures from the Palace Museum, Beijing*, shown in the VMFA galleries from October 18, 2014 through January 19, 2015 marked the initial exhibition portion of the exchange. The deputy director Lou Wei attended the opening at VMFA and presented this successful collaborative project at the US-Sino Cultural Exchange Forum in Washington DC in April 2015. Now, the exhibition of *Fabergé Revealed* marks the first time an American museum's collection is exhibited in the Palace Museum. This rich and enlightening exchange continues at an historic moment—the celebration of the 90th anniversary of the Palace Museum—with the Fabergé Collection filling the galleries of the newly renovated Wumen (Meridian Gate).

This exchange was made possible through the efforts of Ambassador Cui Tiankai of the People's Republic of China and the embassy staff in Washington, D.C., especially Lu Kang, former minister and deputy chief of mission, and Li Hong, the minister counselor from the embassy's Office of Cultural Affairs. Friends of the Virginia Museum of Fine Arts have been incredibly generous, recognizing the critical importance of developing major cultural ties between the

United States and China.

We wish to thank the Palace Museum staff—especially Li Shaoyi, Yuan Hong, Wang Lei, and Wang Siying of the Foreign Affairs department—for their efforts to bring the beauty of the Lillian Thomas Pratt Collection of more than 200 Fabergé and Russian decorative arts objects to Beijing. We also acknowledge the staff of the Forbidden City Publishing House for their contribution to the catalogue and the Palace Museum exhibitions department for their work on the exhibition—Mr. Ji Wei, Deputy Director of the Exhibition Department, Mr. Song Haiyang, Associate Research Fellow, Mr. Wang Bo, Researcher, and Mr. Li Huaiyu, Researcher.

At the VMFA, Barry Shifman, the Sydney and Frances Lewis Family Curator of Decorative Arts from 1890 to the Present, oversees the Lillian Thomas Pratt Collection, assisted on this project by the work of guest curator and Fabergé expert, Géza von Habsburg. The curatorial staff is led by Dr. Michael R. Taylor, Chief Curator and Deputy Director for Art and Education, while Stephen Bonadies, Deputy Director for Facilities and Collections Management, supervises the team that travels with the objects. The Fabergé installation team members—who have traveled with the show since 2012 to five other venues—includes Nancy Nichols, Permanent Collection Registrar; Geoff Strong, Head of Art Handling; Sheila Payaqui, Head of Object Conservation; and Jenn Bridges, Conservation Technician. Organization of the exhibition was handled by Gina Cavallo Collins, Exhibitions Coordinator. Financial management of the project is handled by Hossein Sadid, Chief Financial Officer. Finally, we also recognize Li Jian, E. Rhodes and Leona B. Carpenter Curator of East Asian Art, for the impetus of this partnership, continuous support of all the various exchanges of staff, and for her highly successful curation of the exhibition *Forbidden City: Imperial Treasures from the Palace Museum, Beijing*.

Finally, thank you to all of the staff and supporters of the Virginia Museum of Fine Arts and the Palace Museum for their exceptional work on this catalogue. On behalf of everyone who has had a hand in this collaborative project and historic exchange, I hope you enjoy this beautiful publication that commemorates one of the most important exhibitions in the VMFA's eighty-year, and the Palace Museum's ninety-year, history.

前　言

"法贝热"是精湛工艺、奢侈珠宝以及俄罗斯皇室珍宝的代名词。彼得·卡尔·法贝热（1846～1920年）是俄罗斯一位著名金匠、珠宝首饰匠人、工艺美术设计家。他对俄罗斯传统的珠宝设计和制造工艺进行了大胆的革新。法贝热的作坊以制作精致和巧妙的作品闻名于世，其制作的复活节彩蛋被沙皇俄国和各国皇室皆视为珍品，大部分被世界各地的博物馆或收藏家珍藏。其中，位于美国弗吉尼亚州里士满市的弗吉尼亚美术馆是除俄罗斯本土外，世界最大规模的法贝热珍品集萃地。

弗吉尼亚美术馆成立于1936年，现有藏品超过33000件，藏品历史年限跨越5000年，更因藏品特殊且广泛的展览范围，综合排名位列全美博物馆前十位，是公认的美国顶尖综合性艺术博物馆之一。

今天，故宫博物院与弗吉尼亚美术馆共同为观众呈现法贝热珍品234件（套），这也是美国博物馆的藏品首次大规模在故宫博物院展出。展览通过沙皇的礼物、美轮美奂的装饰艺术、日常的奢华、信仰的力量等四部分系统地展示了法贝热珠宝卓越的理念与工艺，为观众献上别致精美的文化体验。

故宫博物院
弗吉尼亚美术馆

Introduction

"Fabergé" is synonymous with exquisite craftsmanship, luxury jewelry and Russian imperial treasure. Peter Karl Fabergé (1846-1920) was a noted Russian goldsmith, jeweler, and designer of arts and crafts. He brought about bold changes to Russian traditional jewelry design and craftsmanship. The House of Fabergé has been famous worldwide for designing exquisite and ingenious jeweled objects; Fabergé eggs it made as Easter gifts were seen as treasure by imperial families of the Russian Empire and other countries, and most of them have been collected by museums or private collectors all over the world. The Virginia Museum of Fine Arts (VMFA), an art museum in Richmond, Virginia, the United States, houses the largest collection of Fabergé works outside Russia.

Established in 1936, VMFA now has a collection of more than 33,000 works spanning 5,000 years. Because of its special and broad scope of exhibitions, it ranks among the top ten museums and is recognized as one of the top comprehensive art museums in the United States.

Today, the Palace Museum and the Virginia Museum of Fine Arts jointly present 234 pieces (sets) of Fabergé treasure to the audience, for the first time a large collection from a U.S. museum is exhibited in the Palace Museum. The exhibition, which comprises four sections—"Gifts of Tsars", "Majestic Decorative Arts", "Everyday Opulence" and "Power of Faith", is a systematic display of the excellent philosophy and craftsmanship of Fabergé jewelry, offering the audience a fabulous cultural experience.

The Palace Museum
The Virginia Museum of Fine Arts

图版目录

沙皇的礼物

1. 皇家鹈鹕彩蛋 ……………………………………030
2. 沙皇御赐彩蛋
（带有沙皇尼古拉二世的花押）………………036
3. 沙皇御赐彩蛋
（带有皇后亚历山德拉·费奥多罗芙娜的花押）………036
4. 沙皇御赐彩蛋
（带有女大公奥尔加·亚历山德罗芙娜的花押）………038
5. 彼得大帝彩蛋 ……………………………………040
6. 阿列克谢皇储彩蛋 ………………………………044
7. 沙皇御赐彩蛋
（带有皇太后玛利亚·费奥多罗芙娜的花押）………048
8. 沙皇御赐彩蛋
（带有皇太后玛利亚·费奥多罗芙娜的花押）………049
9. 沙皇御赐彩蛋
（带有阿列克谢皇储的花押）…………………049
10. 皇家红十字肖像彩蛋 ……………………………050
11. 戒指盒 ……………………………………………056
12. 袖珍彩蛋吊坠 ……………………………………056
13. 袖珍彩蛋吊坠 ……………………………………056
14. 袖珍彩蛋吊坠 ……………………………………058
15. 袖珍彩蛋吊坠 ……………………………………058
16. 袖珍彩蛋吊坠 ……………………………………058
17. 袖珍彩蛋吊坠 ……………………………………060
18. 袖珍彩蛋吊坠 ……………………………………060
19. 袖珍彩蛋吊坠 ……………………………………060
20. 袖珍彩蛋吊坠
（带有皇后亚历山德拉·费奥多罗芙娜的花押）………061
21. 袖珍彩蛋吊坠 ……………………………………061
22. 袖珍彩蛋吊坠 ……………………………………061
23. 袖珍彩蛋吊坠 ……………………………………064
24. 袖珍彩蛋吊坠 ……………………………………064
25. 袖珍彩蛋吊坠 ……………………………………064
26. 袖珍彩蛋吊坠 ……………………………………065
27. 袖珍彩蛋吊坠 ……………………………………065
28. 袖珍彩蛋吊坠 ……………………………………065
29. 袖珍彩蛋吊坠 ……………………………………068
30. 袖珍彩蛋吊坠 ……………………………………068
31. 袖珍彩蛋吊坠 ……………………………………068
32. 袖珍彩蛋吊坠 ……………………………………070
33. 袖珍彩蛋吊坠 ……………………………………070
34. 袖珍彩蛋吊坠 ……………………………………070

美轮美奂的装饰艺术

35. 钻石别针 …………………………………………074
36. 皇冠胸针 …………………………………………074
37. 皇家钻石胸针 ……………………………………076
38. 十周年纪念胸针 …………………………………078
39. 圣甲虫胸针 ………………………………………080
40. 野玫瑰胸针 ………………………………………080
41. 印章 ………………………………………………082
42. 印章 ………………………………………………082

43. 双幅相框 ……084	69. 相框 ……109
44. 双幅相框 ……084	70. 相框 ……110
45. 双幅相框 ……085	71. 相框 ……110
46. 双幅相框 ……086	72. 相框 ……112
47. 三幅相框 ……087	73. 相框 ……112
48. 三联相框 ……088	74. 相框 ……113
49. 星形相框 ……090	75. 相框 ……113
50. 心形相框 ……091	76. 相框 ……114
51. 皇家银相框 ……093	77. 相框 ……114
52. 十周年纪念相框 ……093	78. 相框 ……116
53. 圣安德烈十字相框 ……094	79. 相框 ……119
54. 皇家柱形相框 ……096	80. 相框 ……119
55. 柱形相框 ……098	81. 相框 ……120
56. 可翻转相框 ……100	82. 相框 ……120
57. 相框 ……102	83. 相框 ……121
58. 相框 ……102	84. 相框 ……122
59. 相框 ……103	85. 相框 ……122
60. 相框 ……103	86. 相框 ……124
61. 相框 ……104	87. 相框 ……125
62. 相框 ……104	88. 相框 ……125
63. 相框 ……105	89. 相框 ……126
64. 相框 ……105	90. 相框 ……126
65. 相框 ……106	91. 相框 ……127
66. 相框 ……106	92. 相框 ……127
67. 相框 ……107	93. 纪念牌匾 ……128
68. 相框 ……108	94. 圆形袖珍肖像 ……130

95. 沙皇台式肖像 130	121. 猪 154
96. 水手雕像 132	122. 象 155
97. 水手雕像礼品盒 134	123. 象 155
98. 鹰 136	124. 象 155
99. 小鸡 136	125. 英国山楂 156
100. 公鸡 137	126. 金莲花 156
101. 公鸡 137	127. 罂粟花 158
102. 犀鸟 137	128. 报春花 159
103. 笼中鸣鸟 138	129. 铃兰 160
104. 鹭 138	130. 金凤花 160
105. 鸵鸟 140	131. 蓝铃花 161
106. 猫头鹰 140	132. 三色堇 162
107. 鸿雁 142	133. 紫苑 162
108. 鸿雁礼品盒 142	134. 郁金香 163
109. 洪堡企鹅 144	135. 马鞭草 164
110. 兔 144	136. 三色堇 165
111. 兔形水罐 144	137. 紫罗兰 166
112. 兔形门铃 146	138. 紫罗兰礼品盒 166
113. 法国斗牛犬 148	
114. 法国斗牛犬 148	**日常的奢华**
115. 法国斗牛犬 150	139. 加冕纪念杯 170
116. 法国斗牛犬 150	140. 袖珍浅杯 172
117. 英国斗牛犬 150	141. 三叶形杯子 174
118. 腊肠犬 152	142. 纺车形杯 175
119. 腊肠犬礼品盒 153	143. 纺车形杯 176
120. 腊肠犬 154	144. 纺车形杯礼品盒 176

145. 纪念版长柄船形酒杯 178	171. 香烟盒 224
146. 长柄船形酒杯 182	172. 香烟盒 224
147. 长柄船形酒杯 182	173. 香烟盒 226
148. 长柄船形酒杯 184	174. 香烟盒 227
149. 袖珍长柄船形酒杯 186	175. 香烟盒 227
150. 赞颂杯 186	176. 图饰珐琅纸牌盒 228
151. 果酒杯 192	177. 象盒 229
152. 广口杯 193	178. 皇家礼品盒 229
153. 杯子 194	179. 阿拉伯文小盒 230
154. 杯子 195	180. 盒子 230
155. 带盖杯 196	181. 盒子 231
156. 带盖啤酒杯 198	182. 盒子 232
157. 带盖杯 200	183. 盒子 233
158. 带盖杯 201	184. 珐琅图饰盒 234
159. 面包和盐之盛器 202	185. 鱼子酱碗 236
160. 盐皿或碗 208	186. 鼻烟壶 238
161. 盘 208	187. 第一次世界大战纪念烟灰缸 238
162. 果子露杯、托盘和勺 212	188. 墨水池 239
163. 咖啡具三件套 214	189. 花瓶 240
164. 小咖啡勺 218	190. 硬币花瓶 242
165. 公用匙 219	191. 书签 243
166. 第一次世界大战纪念碟 220	192. 地球仪 243
167. 糖果盒 221	193. 项坠 244
168. 糖果盒 221	194. 沙皇孤儿院徽章 （带有皇后亚历山德拉·费奥多罗芙娜的花押）...... 244
169. 糖果盒 222	
170. 糖果盒 222	195. 万年历 245

196. 伞柄	246
197. 伞柄	246
198. 伞柄	248
199. 伞柄	248
200. 伞柄	250
201. 伞柄	250
202. 伞柄	251
203. 伞柄	251
204. 伞柄	252
205. 伞柄	252
206. 伞柄	254
207. 伞柄	254
208. 伞柄	255
209. 伞柄	255
210. 伞柄	256
211. 伞柄礼盒	256
212. 伞柄	258
213. 手杖柄	258
214. 手杖柄	260
215. 手杖柄	260
216. 手杖柄	262
217. 手杖柄	262
218. 手杖柄	264
219. 手杖柄	265
220. 手杖柄	265

信仰的力量

221.《十二门徒圣像》	268
222.《艾弗斯卡亚圣母》	270
223.《抹大拉的圣玛利亚、奇迹创造者圣尼古拉、圣亚历山大·涅夫斯基大公》	272
224.《圣亚历山德拉公主与奇迹创造者圣尼古拉》	274
225.《全能的基督》	278
226.《容光焕发的基督》	280
227.《奇迹创造者圣尼古拉》	282
228.《奇迹创造者圣尼古拉》	284
229.《全能的基督》	285
230.《艾弗斯卡亚圣母》	286
231.《圣容显现、圣伊丽莎白、拉多聂兹的圣谢尔盖三联画》	288
232.《喀山圣母、圣亚力山大·涅夫斯基大公、抹大拉的圣玛利亚三联画》	290
233.《耶稣复活三联画》	292
234. 圣乔治屠龙吊坠	294

List of Plates

Gifts of Tsars

1. Imperial Pelican Easter Egg 032
2. Imperial Presentation Porcelain Easter Egg
   with the Cypher of Tsar Nicholas II 036
3. Imperial Presentation Porcelain Easter Egg
   with the Cypher of Tsaritsa Alexandra Feodorovna 036
4. Imperial Presentation Porcelain Easter Egg
   with the Cypher of Grand Duchess Olga Aleksandrovna 038
5. Imperial Peter the Great Easter Egg 042
6. Imperial Tsesarevich Easter Egg 046
7. Imperial Presentation Porcelain Easter Egg
   with the Cypher of Dowager Empress Maria Feodorovna 048
8. Imperial Presentation Porcelain Easter Egg
   with the Cypher of Dowager Empress Maria Feodorovna 049
9. Imperial Presentation Porcelain Easter Egg
   with the Cypher of Tsesarevich Aleksei Nikolaevich 049
10. Imperial Red Cross Easter Egg with Portraits 052
11. Ring Box 056
12. Miniature Easter Egg Pendant 056
13. Miniature Easter Egg Pendant 056
14. Miniature Easter Egg Pendant 058
15. Miniature Easter Egg Pendant 058
16. Miniature Easter Egg Pendant 058
17. Miniature Easter Egg Pendant 060
18. Miniature Easter Egg Pendant 060
19. Miniature Easter Egg Pendant 060
20. Miniature Easter Egg Pendant with Cypher
    for Tsaritsa Alexandra Feodorovna 061
21. Miniature Easter Egg Pendant 061
22. Miniature Easter Egg Pendant 061
23. Miniature Easter Egg Pendant 064
24. Miniature Easter Egg Pendant 064
25. Miniature Easter Egg Pendant 064
26. Miniature Easter Egg Pendant 065
27. Miniature Easter Egg Pendant 065
28. Miniature Easter Egg Pendant 065
29. Miniature Easter Egg Pendant 068
30. Miniature Easter Egg Pendant 068
31. Miniature Easter Egg Pendant 068
32. Miniature Easter Egg Pendant 070
33. Miniature Easter Egg Pendant 070
34. Miniature Easter Egg Pendant 070

Majestic Decoratire Arts

35. Diamond Pin 074
36. Crown Brooch 074
37. Imperial Diamond Brooch 076
38. Tenth-Anniversary Brooch 078
39. Scarab Brooch 080
40. Wild Rose Brooch 080
41. Seal 082
42. Seal 082
43. Double Frame 084
44. Double Frame 084
45. Double Frame 085
46. Double Frame 086

47. Triple Frame 087	75. Frame 113
48. Triptych Frame 088	76. Frame 114
49. Star Frame 090	77. Frame 114
50. Heart Frame 091	78. Frame 116
51. Imperial Silver Frame 093	79. Frame 119
52. Tenth-Anniversary Frame 093	80. Frame 119
53. Cross of St. Andrew Frame 094	81. Frame 120
54. Imperial Column Portrait Frame 096	82. Frame 120
55. Column Frame 098	83. Frame 121
56. Reversible Frame 100	84. Frame 122
57. Frame 102	85. Frame 122
58. Frame 102	86. Frame 124
59. Frame 103	87. Frame 125
60. Frame 103	88. Frame 125
61. Frame 104	89. Frame 126
62. Frame 104	90. Frame 126
63. Frame 105	91. Frame 127
64. Frame 105	92. Frame 127
65. Frame 106	93. Commemorative Plaque 128
66. Frame 106	94. Circular Miniature 130
67. Frame 107	95. Imperial Table Portrait 130
68. Frame 108	96. Statuette of a Sailor 132
69. Frame 109	97. Presentation Box for Statuette of a Sailor 134
70. Frame 110	98. Eagle 136
71. Frame 110	99. Chick 136
72. Frame 112	100. Rooster 137
73. Frame 112	101. Rooster 137
74. Frame 113	102. Hornbill 137

103. Songbird in a Cage 138	**Everyday Opulence**
104. Heron 138	139. Coronation Beaker 170
105. Ostrich 140	140. Miniature Tazza 172
106. Owl 140	141. Trefoil Cup 174
107. Chinese Swan Goose 142	142. Charka 175
108. Presentation Box for Chinese Swan Goose ... 142	143. Charka 176
109. Humboldt Penguin 144	144. Presentation Box for Charka 176
110. Rabbit 144	145. Monumental Kovsh 180
111. Rabbit Pitcher 144	146. Kovsh 182
112. Rabbit Bell Push 146	147. Kovsh 182
113. French Bulldog 148	148. Kovsh 184
114. French Bulldog 148	149. Miniature Kovsh 186
115. French Bulldog 150	150. Loving Cup 186
116. French Bulldog 150	151. Bratina 192
117. English Bulldog 150	152. Beaker 193
118. Dachshund 152	153. Cup 194
119. Presentation Box for Dachshund 153	154. Cup 195
120. Dachshund 154	155. Cup and Cover 196
121. Pig 154	156. Tankard and Cover 198
122. Elephant 155	157. Cup and Cover 200
123. Elephant 155	158. Cup and Cover 201
124. Elephant 155	159. Bread-and-Salt Dish 204
125. English Hawthorn 156	160. Salt Cellar or Bowl 208
126. Globeflowers 156	161. Plate 208
127. Poppy 158	162. Sherbet Cup, Saucer, and Spoon 212
128. Primrose 159	163. Three-Piece Coffee Set 214
129. Lilies of the Valley 160	164. Demitasse Spoons 218
130. Buttercup 160	165. Serving Spoon 219
131. Bluebells 161	166. World War I Dish 220
132. Pansies 162	167. Bonbonnière 221
133. Aster 162	168. Bonbonnière 221
134. Tulip 163	169. Bonbonnière 222
135. Verbena 164	170. Bonbonnière 222
136. Pansy 165	171. Cigarette Case 224
137. Violet 166	172. Cigarette Case 224
138. Presentation Box for Violet 166	173. Cigarette Case 226

174. Cigarette Case	227
175. Cigarette Case	227
176. Pictorial Enamel Cardcase	228
177. Elephant Box	229
178. Imperial Presentation Box	229
179. Box with Arabic Inscription	230
180. Box	230
181. Box	231
182. Box	232
183. Box	233
184. Enamel Pictorial Box	234
185. Caviar Bowl	237
186. Snuffbox	238
187. World War I Ashtray	238
188. Inkwell	239
189. Vase	240
190. Coin Vase	242
191. Bookmark	243
192. Terrestrial Globe	243
193. Locket	244
194. Badge of the Imperial Orphanage with the Cypher of Tsaritsa Alexandra Feodorovna	244
195. Perpetual Calendar	245
196. Parasol Handle	246
197. Parasol Handle	246
198. Parasol Handle	248
199. Parasol Handle	248
200. Parasol Handle	250
201. Parasol Handle	250
202. Parasol Handle	251
203. Parasol Handle	251
204. Parasol Handle	252
205. Parasol Handle	252
206. Parasol Handle	254
207. Parasol Handle	254
208. Parasol Handle	255
209. Parasol Handle	255
210. Parasol Handle	256
211. Presentation Box for Parasol Handle	256
212. Parasol Handle	258
213. Cane Handle	258
214. Cane Handle	260
215. Cane Handle	260
216. Cane Handle	262
217. Cane Handle	262
218. Cane Handle	264
219. Cane Handle	265
220. Cane Handle	265

Power of Faith

221. Icon of Twelve Saints	268
222. The Iverskaya Mother of God	270
223. St. Mary Magdalene, St. Nicholas the Miracle Worker, St. Prince Aleksandr Nevskii	272
224. Princess St. Alexandra and St. Nicholas the Miracle Worker Diptych	274
225. Christ Pantocrator	278
226. Christ Transfigured	280
227. St. Nicholas the Miracle Worker	282
228. St. Nicholas the Miracle Worker	284
229. Christ Pantocrator	285
230. The Iverskaya Mother of God	286
231. The Transfiguration, St. Elizabeth, and St. Sergius of Radonezh Triptych	288
232. The Holy Virgin of Kazan, St. Prince Aleksandr Nevskii, St. Mary Magdalene Triptych	290
233. The Resurrection Triptych	292
234. Pendant of St. George Slaying the Dragon	294

图 版
Plates

沙皇的礼物
Gifts of Tsars

  法贝热最杰出的成就是为俄罗斯沙皇制作的 50 枚独具魅力的复活节彩蛋，其灵感来源于复活节赠送彩蛋的俄罗斯传统习俗。1885 年，他接到了第一笔订单：沙皇亚历山大三世命令他为皇后制作一枚复活节彩蛋。由于这枚彩蛋构思巧妙，制作精美，1886 年法贝热被沙皇封为"皇家御用珠宝师"。从 1894 年开始，沙皇尼古拉二世继续沿袭其父传统，每年复活节都会给母亲玛利亚·费奥多罗芙娜和妻子亚历山德拉·费奥多罗芙娜赠送彩蛋。这些蛋雕以珍贵的金属或是半宝石混合珐琅与宝石来装饰。"法贝热彩蛋"后来成为奢侈品的代名词，并且被认为是珠宝艺术的经典之作。

  The greatest achievement of Fabergé is the fifty Easter eggs—each of them has a unique style—which he made for the last two tsars, Nicholas II and Alexander III, for which he drew inspiration from the Russian tradition of gifting decorated eggs to celebrate Easter. In 1885, Fabergé received his first order for an Easter egg: Tsar Alexander III commissioned Fabergé to make an Easter egg as a gift for his wife. His first Easter egg was so ingeniously designed and exquisitely made that he was appointed the "Jeweler to the Imperial Court" by the tsar in 1886. From 1894 onwards, Tsar Nicholas II continued his father's tradition and each year gave his mother Maria Feodorovna and his wife Alexandra Feodorovna Fabergé eggs as Easter gifts. These Easter eggs were made with precious metals or semiprecious stones decorated with enamels and gemstones. Fabergé eggs later became synonymous with luxuries and have since been recognized as masterpieces of jewelry arts.

1

皇家鹈鹕彩蛋

1897 年
法贝热公司
工匠大师：米哈伊尔·佩尔欣
黄金、钻石、珐琅、珍珠、猛犸象牙、水彩、玻璃
彩蛋：高 101.6 毫米　直径 53.975 毫米
支架：高 63.5 毫米　宽 66.675 毫米
弗吉尼亚美术馆藏
莉莲·托马斯·普拉特遗赠

沙皇鹈鹕彩蛋是沙皇尼古拉二世 1898 年献给其母亲玛利亚·费奥多罗芙娜皇太后的复活节礼物，同时也是庆祝沙皇亚历山大一世（在位时间为 1801～1825 年）之母玛利亚·费奥多罗芙娜创建慈善机构 100 周年的纪念物。彩蛋顶部立着一只鹈鹕，因而得名鹈鹕彩蛋。鹈鹕站在窝里，守护着窝中的三只雏鸟。鹈鹕象征着皇太后本人，也被人们普遍认为是基督牺牲自我的母爱象征：因为鹈鹕会以自己的血肉喂食和抚养小鹈鹕。

这只红金色彩蛋展开后，是八幅尺寸依次变小且边缘镶有珍珠的椭圆形相框。每个相框内都镶有宫廷画家约翰尼斯·津格拉夫所绘的微型象牙画，依次描绘了玛利亚·费奥多罗芙娜资助的孤儿院和教育机构。在第四、五幅画框中间设计有一个扁平的黄金支架，作为八幅画的支撑，支架上雕刻着象征科学和艺术的标志。

微型画（从左到右）：圣彼得堡伊丽莎白学院（1808 年），圣彼得堡尼古拉孤儿院（1837 年），圣彼得堡叶卡捷琳娜学院（1798 年），圣彼得堡帕夫洛夫斯基学院（1853 年），圣彼得堡斯莫尔尼学院（1764 年），圣彼得堡爱国学院（1827 年），圣彼得堡克仙妮学院（1894 年），以及莫斯科尼古拉孤儿院（1837 年）。

1898 年至 1917 年期间，沙皇鹈鹕彩蛋一直陈列在圣彼得堡玛利亚·费奥多罗芙娜皇太后的阿尼奇科夫宫里。1902 年 3 月，彩蛋曾在皇后亚历山德拉赞助的圣彼得堡冯德威斯男爵公馆的慈善展览上展出。1917 年，临时政府下令将其没收，转移至莫斯科的克里姆林宫军械库里由其保管。1930 年，古董局（列宁时期向西方售卖国家财产的专设机构）以 1000 卢布的价格将其出售给阿曼德·哈默。在 1930 年至 1936 年到 1938 年的某个时间，彩蛋一直陈列在纽约的阿曼德·哈默画廊。之后在 1936 至 1938 年的某个时间，这件皇家宝物被莉莲·托马斯·普拉特购得。

1

Imperial Pelican Easter Egg

1897
Fabergé firm
Workmaster: Mikhail Perkhin
Gold, diamonds, enamel, pearls, Mammoth ivory, watercolor, glass
Egg: 4" H × 2 1/8" Dia.
Stand: 2 1/2" H × 2 5/8" W
Virginia Museum of Fine Arts
Bequest of Lillian Thomas Pratt

The *Imperial Pelican Easter Egg* was presented by Tsar Nicholas II to his mother, Dowager Empress Maria Feodorovna in 1898. The egg commemorates the 100th anniversary of the founding of the charities by Maria Feodorovna, the mother of Tsar Alexander I (reigned 1801-1825). The pelican that gives the egg its name balances at the top, standing in a nest that holds three chicks. The pelican was a personal symbol of the dowager empress, as well as a recognized symbol of maternal care that recalls the sacrifice of Christ: the pelican tears her flesh so that her children may feed and live.

This red-gold egg unfolds into eight oval frames graduated in size and rimmed with pearls. Each frame contains a miniature on ivory by the imperial court painter Johannes Zehngraf of orphanages and educational institutions of which Maria Feodorovna was patroness. The frames are held upright by a flat gold stand, engraved with symbols of science and the arts, anchored between the fourth and fifth frames.

Miniatures (left to right): The Elisavetinskii Institute, St. Petersburg (1808), The Nikolaevskii Orphanage, St. Petersburg (1837), The Ekaterininski Institute, St. Petersburg (1798), The Pavlovskii Institute, St. Petersburg (1853), The Smolnyi Institute, St. Petersburg (1764), The Patriotic Institute, St. Petersburg (1827), The Kseniinski Institute, St. Petersburg (1894), and the Nikolaevskii Orphanage, Moscow (1837).

The Imperial Pelican Easter Egg was displayed in the Dowager Empress Maria Feodorovna's Anichkov Palace in St. Petersburg from 1898 to 1917. It was on view in March 1902 at the charity exhibition held under the patronage of Empress Alexandra at the mansion of Baron von Dervis in St. Petersburg. In 1917, the egg was confiscated by order of the Provisional Government and sent to the Kremlin Armory in Moscow for safekeeping. Sold by Antikvariat (Lenin's bureau established to sell state treasures to the West) to Armand Hammer for 1,000 rubles in 1930, it was with the Hammer Galleries in New York City between 1930 and 1936 to 1938. Lillian Thomas Pratt acquired this imperial treasure sometime between 1936 and 1938.

2

沙皇御赐彩蛋

（带有沙皇尼古拉二世的花押）

约 1900 年
沙皇瓷器厂，圣彼得堡
瓷、珐琅、镀金
高 85.725 毫米　宽 69.85 毫米
弗吉尼亚美术馆藏
爱丽丝·道奇小姐赠送，以纪念其父亲
亨利·珀西瓦尔·道奇
（美国驻巴拿马公使，1911～1913 年）
和母亲玛格利特·亚当斯·道奇

2

Imperial Presentation Porcelain Easter Egg with the Cypher of Tsar Nicholas II

ca. 1900
Imperial Porcelain Manufactory,
St. Petersburg
Porcelain, enamel, gilding
3 3/8" H × 2 3/4" W
Virginia Museum of Fine Arts
Gift of Miss Alice Dodge in memory of her parents,
Henry Percival Dodge, American Minister to Panama,
1911-13, and Margaret Adams Dodge

3

沙皇御赐彩蛋

（带有皇后亚历山德拉·费奥多罗芙娜的花押）

约 1900 年
沙皇瓷器厂，圣彼得堡
瓷、珐琅、镀金
高 88.9 毫米　直径 69.85 毫米
弗吉尼亚美术馆藏
莉莲·托马斯·普拉特遗赠，

3

Imperial Presentation Porcelain Easter Egg with the Cypher of Tsaritsa Alexandra Feodorovna

ca. 1900
Imperial Porcelain Manufactory,
St. Petersburg
Porcelain, enamel, gilding
3 1/2" H × 2 3/4" Dia.
Virginia Museum of Fine Arts
Bequest of Lillian Thomas Pratt

4

沙皇御赐彩蛋

（带有女大公奥尔加·亚历山德罗芙娜的花押）

约 1900 年
沙皇瓷器厂，圣彼得堡
瓷、珐琅、镀金
高 88.9 毫米　宽 76.2 毫米
弗吉尼亚美术馆藏
爱丽丝·道奇小姐赠送，以纪念其父亲
亨利·珀西瓦尔·道奇
(美国驻巴拿马公使，1911 ～ 1913 年) 和
母亲玛格利特·亚当斯·道奇

4

Imperial Presentation Porcelain Easter Egg with the Cypher of Grand Duchess Olga Aleksandrovna

ca. 1900
Imperial Porcelain Manufactory,
St. Petersburg
Porcelain, enamel, gilding
3 1/2" H × 3" W
Virginia Museum of Fine Arts,
Gift of Miss Alice Dodge in memory of her parents,
Henry Percival Dodge, American Minister to Panama, 1911-13,
and Margaret Adams Dodge

5

彼得大帝彩蛋

1903 年
法贝热公司
工匠大师：米哈伊尔·佩尔欣
黄金、铂、银镀金、钻石、红宝石、珐琅、水彩、猛犸象牙、水晶、铜鎏金、蓝宝石
彩蛋：高 120.65 毫米　直径 79.375 毫米
微型画：高 47.625 毫米　宽 69.85 毫米
支架：高 77.788 毫米　宽 69.85 毫米
弗吉尼亚美术馆藏
莉莲·托马斯·普拉特遗赠

彼得大帝彩蛋是沙皇尼古拉二世 1903 年送给皇后亚历山德拉·费奥多罗芙娜的礼物，作为圣彼得堡建城 200 周年的纪念。彩蛋的一侧镶嵌有一幅彼得大帝的微型画像和他的小木屋，即该市第一座建筑，旁边标注着年份"1703"年；另一侧则是尼古拉二世的画像和皇家宫殿冬宫的微型画，旁边标注着年份"1903"年。彩蛋底部则是双头鹰徽记和圣乔治屠龙的微型画。

打开蛋壳，一幅仿造艾蒂安·莫里斯·法尔科内特的名作——彼得大帝骑马像制作的微型雕塑缓缓升起。在这令人称奇的雕像中，沙皇一身罗马英雄的打扮，横跨一匹扬起前蹄的骏马。原作由叶卡捷琳娜大帝于 1782 年委托制作，以示向著名的彼得大帝——她的前任沙皇致敬。雕像的底座为一块重达 1000 吨的花岗岩，底座上刻有拉丁文和俄罗斯文："献给彼得一世，叶卡捷琳娜二世。"这座雕像后来被称为"青铜骑士"，因为它是俄罗斯最伟大的诗人亚历山大·普希金的诗歌《青铜骑士》的灵感源泉。

法贝热以艾尔米塔什博物馆里见到的一只 18 世纪中期的蛋形日用小件（用来盛放针线或其他小物品的器皿）器型为模板，创作了彼得大帝彩蛋。弗吉尼亚美术馆收藏的这枚彩蛋淋漓尽致地展现了法贝热对彩金工艺的娴熟应用。

1917 年，临时政府下令将彼得大帝彩蛋没收，并送至莫斯科克里姆林宫军械库保管。1933 年，古董局（列宁时期向西方售卖国家财产的专设机构）将彩蛋出售给一位匿名买家；20 世纪 30 年代至 1942 年期间，彩蛋成为亚历山大·谢弗的俄罗斯皇家艺术珍品收藏的一部分；1942 年被莉莲·托马斯·普拉特在纽约购得。

Imperial Peter the Great Easter Egg

1903
Fabergé firm
Workmaster: Mikhail Perkhin
Egg: gold, platinum, silver gilt, diamonds, rubies, enamel, watercolor,
Mammoth ivory, rock crystal, statue: gilt bronze, sapphire
Egg: 4 3/4" H × 3 1/8" Dia.
Miniature: 1 7/8" H × 2 3/4" W
Stand: 3 1/16" H × 2 3/4" W
Virginia Museum of Fine Arts
Bequest of Lillian Thomas Pratt

This *Imperial Peter the Great Easter Egg*, which commemorates the 200th anniversary of the founding of St. Petersburg, was presented by Tsar Nicholas II to his wife Tsaritsa Alexandra Feodorovna in 1903. On one side of the egg, a miniature portrait of Peter the Great appears with the date 1703 and a picture of his humble log hut, the first structure in the city. On the other side of the egg, a portrait of Nicholas II appears with the date 1903, and complements a miniature painting of the Winter Palace, the official imperial residence. On the base of the egg there is the double-headed eagle and a miniature of St. George and the dragon.

When the egg opens, a miniature replica of Étienne-Maurice Falconet's famous statue of Peter the Great on horseback rises out of the shell. The monumental statue that inspired this surprise depicts the tsar dressed as a Roman hero astride a rearing horse. Commissioned by Catherine the Great in 1782 as a tribute to her famous predecessor, the monument stands on top of a thousand-ton granite base inscribed in Russian and Latin, *To Peter the First [from] Catherine the Second*. This statute has come to be known as the "Bronze Horseman" because it inspired a poem with that title by Aleksandr Pushkin, Russia's greatest poet.

Fabergé used a mid-18th-century egg-shaped *nécessaire* (a container for sewing implements or other small objects) found in the Hermitage as a model for the *Imperial Peter the Great Easter Egg*. The VMFA's imperial egg provides a striking example of Fabergé's accomplished use of multicolored gold.

In 1917, the *Imperial Peter the Great Easter Egg* was confiscated by order of the Provisional Government and sent to the Kremlin Armory in Moscow for safe-keeping. Sold by Antikvariat (Lenin's bureau established to sell state treasures to the West) to an anonymous buyer in 1933, it was acquired by The Schaffer Collection of Russian Imperial Art Treasures sometime between the 1930s and 1942—and was bought in New York City by Lillian Thomas Pratt in 1942.

阿列克谢皇储彩蛋

1912 年
法贝热公司
工匠大师：亨瑞克·魏格斯特姆
彩蛋：青金石、黄金、钻石，画框：铂、青金石、钻石、水晶
微型画为复制品，原作为猛犸象牙水彩画
彩蛋：高 123.825 毫米　直径 90.488 毫米
微型画：高 95.25 毫米　宽 60.325 毫米
支架：高 66.675 毫米　宽 104.775 毫米
弗吉尼亚美术馆藏
莉莲·托马斯·普拉特遗赠

这枚沙皇皇储阿列克谢彩蛋是沙皇尼古拉二世于 1912 年赠予皇后亚历山德拉·费奥多罗芙娜的礼物。该彩蛋由六块青金石制成，壳体饰有黄金雕琢的双头鹰、双翼女像柱、华盖、涡卷、花篮和浪花图案，用来掩盖接缝。这些黄金装饰表明，法贝热热衷于改造 17 世纪晚期法国艺术家让·贝郎的版画设计。彩蛋底部镶有一大枚独粒钻石，而在顶部镶有一枚盘形钻石（平整而纤薄的钻石），顶部还刻有西里尔字母"AF"（即亚历山德拉·费奥多罗芙娜的缩写）和创作时间"1912"年。

彩蛋内令人称奇之处在于一尊立于青金石基座的镶钻双头鹰。基座上是一幅绘于象牙之上的皇储阿列克谢（沙皇王位继承人）肖像。亚历山德拉皇后于 1904 年 8 月 12 日诞下一子，名唤阿列克谢。由于阿列克谢王子天生患有血友病（一种罕见的血液疾病），在他短暂的一生中给双亲带来了极大的忧虑。

1912 年至 1917 年间，这枚沙皇皇储阿列克谢彩蛋珍藏于圣彼得堡亚历山大宫皇后的淡紫色房间。1917 年，经临时政府下令，该枚彩蛋被转移至克里姆林宫军械库博物馆保管。1930 年，古董局（列宁时期向西方售卖国家财产的专设机构）以 8000 卢布的价格将其卖给阿曼德·哈默，后者于 1934 年又将这枚著名的皇家彩蛋售予莉莲·托马斯·普拉特。参照某张早期相片得知，该彩蛋的底座是原配遗失之后的现代替代品。

Imperial Tsesarevich Easter Egg

1912
Fabergé firm
Workmaster: Henrik Wigström
Egg: lapis lazuli, gold, diamonds,
Picture frame: platinum, lapis lazuli, diamonds, rock crystal,
(reproduction miniature) original watercolor on Mammoth ivory
Egg: 4 7/8" H × 3 9/16" Dia.
Miniature: 3 3/4" H × 2 3/8" W
Stand: 2 5/8" H × 4 1/8" W
Virginia Museum of Fine Arts
Bequest of Lillian Thomas Pratt

The *Imperial Tsesarevich Easter Egg* was presented by Tsar Nicholas II to his wife Tsaritsa Alexandra Feodorovna in 1912. The egg has six lapis lazuli sections with applied gold decoration comprised of double-headed eagles, winged caryatids, hanging canopies, scrolls, flower baskets, and sprays that conceal the joints. This gold decoration shows Fabergé's interest in adapting designs from late 17th century engraved prints by the French artist Jean Bérain. The egg is set with a large solitaire diamond at the base, and a table diamond (a thin, flat diamond) on top over the Cyrillic monogram AF (for Alexandra Feodorovna) and the date 1912.

The surprise found inside the egg is a diamond-set, double-headed eagle standing on a lapis lazuli pedestal. On the pedestal is a portrait on ivory of the Tsesarevich (heir to the Russian throne). Empress Alexandra gave birth to a son called Aleksei on August 12, 1904. Born with hemophilia, a rare blood disease, Aleksei's illness caused both his parents great distress during his short lifetime.

During the period 1912 to 1917, the *Imperial Tsesarevich Easter Egg* was in the Empress's Mauve Room at the Alexander Palace, St. Petersburg. In 1917, it was transferred by order of the Provisional Government to the Kremlin Armory in Moscow for safekeeping. The egg was later sold by Antikvariat (Lenin's bureau established to sell state treasures to the West) to Armand Hammer for 8,000 rubles, who owned it from 1930 to 1934. Lillian Thomas Pratt acquired this well-known imperial treasure in 1934. Based on an early photograph, the stand for the egg is a modern replacement of the lost original.

7

沙皇御赐彩蛋
（带有皇太后玛利亚·费奥多罗芙娜的花押）

1914 ~ 1916 年
沙皇瓷器厂，圣彼得堡
瓷、珐琅、镀金
高 66.675 毫米　直径 63.5 毫米
弗吉尼亚美术馆藏
莉莲·托马斯·普拉特遗赠

7

Imperial Presentation Porcelain Easter Egg with the Cypher of Dowager Empress Maria Feodorovna

1914–1916
Imperial Porcelain Manufactory,
St. Petersburg
Porcelain, enamel, gilding
2 5/8" H × 2 1/4" Dia.
Virginia Museum of Fine Arts
Bequest of Lillian Thomas Pratt

8

沙皇御赐彩蛋
（带有皇太后玛利亚·费奥多罗芙娜的花押）

1914 ~ 1916 年
沙皇瓷器厂，圣彼得堡
瓷、珐琅、镀金、银
高 66.675 毫米　直径 57.15 毫米
弗吉尼亚美术馆藏
莉莲·托马斯·普拉特遗赠

8

Imperial Presentation Porcelain Easter Egg with the Cypher of Dowager Empress Maria Feodorovna

1914–1916
Imperial Porcelain Manufactory,
St. Petersburg
Porcelain, enamel, gilding, silver
2 5/8" H × 2 1/4" Dia.
Virginia Museum of Fine Arts
Bequest of Lillian Thomas Pratt

9

沙皇御赐彩蛋
（带有阿列克谢皇储的花押）

1914 ~ 1916 年
沙皇瓷器厂，圣彼得堡
瓷、镀金
高 66.675 毫米　直径 53.975 毫米
弗吉尼亚美术馆藏
莉莲·托马斯·普拉特遗赠

9

Imperial Presentation Porcelain Easter Egg with the Cypher of Tsesarevich Aleksei Nikolaevich

1914–1916
Imperial Porcelain Manufactory,
St. Petersburg
Porcelain, gilding
2 5/8" H × 2 1/8" Dia.
Virginia Museum of Fine Arts
Bequest of Lillian Thomas Pratt

10

皇家红十字肖像彩蛋

1915 年
法贝热公司
工匠大师：亨瑞克·魏格斯特姆
黄金、银镀金、珐琅、螺钿、猛犸象牙、水彩、天鹅绒、玻璃
肖像（从左至右）：沙皇的姐姐奥尔加女大公、沙皇的大女儿奥尔加女大公、
皇后亚历山德拉·费奥多罗芙娜、沙皇的二女儿塔季扬娜女大公、
沙皇的堂妹玛利亚·帕夫洛夫娜女大公
彩蛋：高 76.2 毫米　直径 60.325 毫米
微型画：高 49.213 毫米　宽 190.5 毫米　厚 6.35 毫米
支架：高 58.738 毫米　宽 58.738 毫米
弗吉尼亚美术馆藏
莉莲·托马斯·普拉特遗赠

　　这枚皇家红十字彩蛋，是沙皇尼古拉二世于 1915 年献给母亲玛利亚·费奥多罗芙娜皇太后的礼物，以此颂扬她任俄国红十字会主席一职所做的贡献。彩蛋内有五片白色玑镂珐琅，各自饰有不同图案。居中的一片，饰以两个鲜艳的红十字，上有斯拉夫语镀金铭文："人为朋友舍命，是爱心最大的表现。"摘自《约翰福音》15 章 13 节。彩蛋顶部饰有皇冠与玛利亚·费奥多罗芙娜姓名首字母 MF 的西里尔文交织图案。

　　这枚彩蛋令人称奇的地方，在于内含一件可拆卸围屏。白色玑镂珐琅嵌板之上为一组肖像，嵌板之后镶有螺钿与肖像人物的姓名首字母交织图案。穿戴成慈悲修女模样的肖像人物，分别是年轻的皇后亚历山德拉·费奥多罗芙娜，她最大的两个女儿奥尔加与塔季扬娜，以及另外两位与皇后关系密切的女性亲戚。玛利亚·费奥多罗芙娜的儿媳亚历山德拉·费奥多罗芙娜和她最大的两个女儿曾在皇后于亚历山大宫主持开设的临时医院里帮助受伤的一战士兵。玛利亚·帕夫洛夫娜女大公是帕维尔·亚历山德罗大公的女儿，1908 年嫁于瑞典威廉王子。1913 年与其离婚后，她重返俄国，在一战中担任一名护士。

　　这枚皇家红十字彩蛋很有可能收藏于玛利亚·费奥多罗芙娜位于圣彼得堡的阿尼奇科夫宫，直到 1917 年被临时政府下令没收充公，后送往莫斯科克里姆林宫武器军械库保管。1930 年，古董局（列宁时期向西方售卖国家财产的专设机构）将其出售给阿曼德·哈默；1930 年至 1933 年间，这枚彩蛋收藏于纽约哈默画廊。1933 年，莉莲·托马斯·普拉特购入了这枚彩蛋。

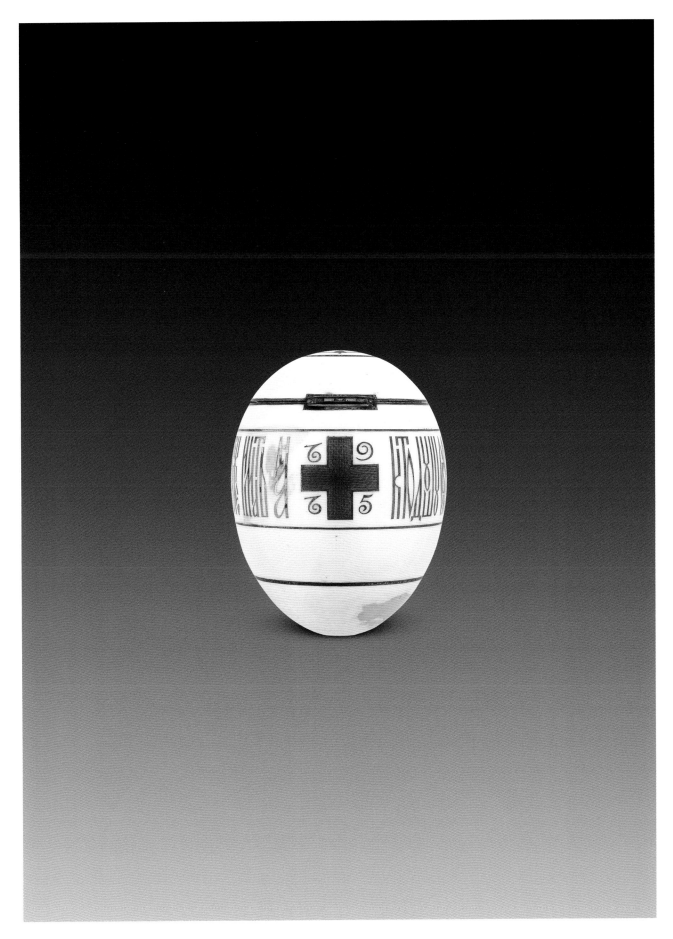

Imperial Red Cross Easter Egg with Portraits

1915
Fabergé firm
Workmaster: Henrik Wigström
Gold, silver gilt, enamel, mother-of-pearl, watercolor on Mammoth ivory, velvet, glass
Portraits: (left to right): Grand Duchess Olga, the tsar's sister;
Grand Duchess Olga, eldest daughter of the tsar; Tsaritsa Alexandra Feodorovna;
Grand Duchess Tatiana, the tsar's second daughter;
and the Grand Duchess Maria Pavlovna, the tsar's first cousin
Egg: 3" H × 2 3/8" Dia.
Miniatures: 1 15/16" H × 7 1/2" W × 1/4" D
Stand: 2 5/16" H × 2 5/16" W
Virginia Museum of Fine Arts
Bequest of Lillian Thomas Pratt

The *Imperial Red Cross Easter Egg*, presented by Tsar Nicholas II in 1915 to his mother, Dowager Empress Maria Feodorovna, is a tribute to her service as president of the Russian Red Cross. The egg has five bands of white guilloché enamel, each with a different pattern. The central band, ornamented with two brilliant-red crosses, has a gilded Slavonic inscription: *Greater love hath no man than this, that a man lay down his life for his friends* (John 15:13). The top of the egg has the crown and Cyrillic monogram MF for Maria Feodorovna.

The egg's surprise is a removable folding screen with portraits within panels of white guilloché enamel. The backs of the panels are of mother-of-pearl and bear the monograms of the sitters who are dressed as Sisters of Mercy. They include the young empress Alexandra Feodorovna, her two eldest daughters, Olga and Tatiana, and two close female relatives. Maria Feodorovna's daughter-in-law Alexandra Feodorovna and her two eldest daughters helped wounded and dying World War I soldiers in a hospital organized by the tsaritsa at the Alexander Palace. Grand Duchess Maria Pavlovna, daughter of Grand Duke Pavel Aleksandrovich, was married to Prince Wilhelm of Sweden in 1908, with whom she divorced in 1913. She returned to Russia and served as a nurse during World War I.

The *Imperial Red Cross Easter Egg* was probably in Maria Feodorovna's Anichkov Palace in St. Petersburg until 1917. That year it was confiscated by order of the Provisional Government and sent to the Kremlin Armory in Moscow for safekeeping. Sold by Antikvariat (Lenin's bureau established to sell state treasures to the West) to Armand Hammer in 1930, it was with the Hammer Galleries in New York City between 1930 and 1933. Lillian Thomas Pratt acquired the egg in 1933.

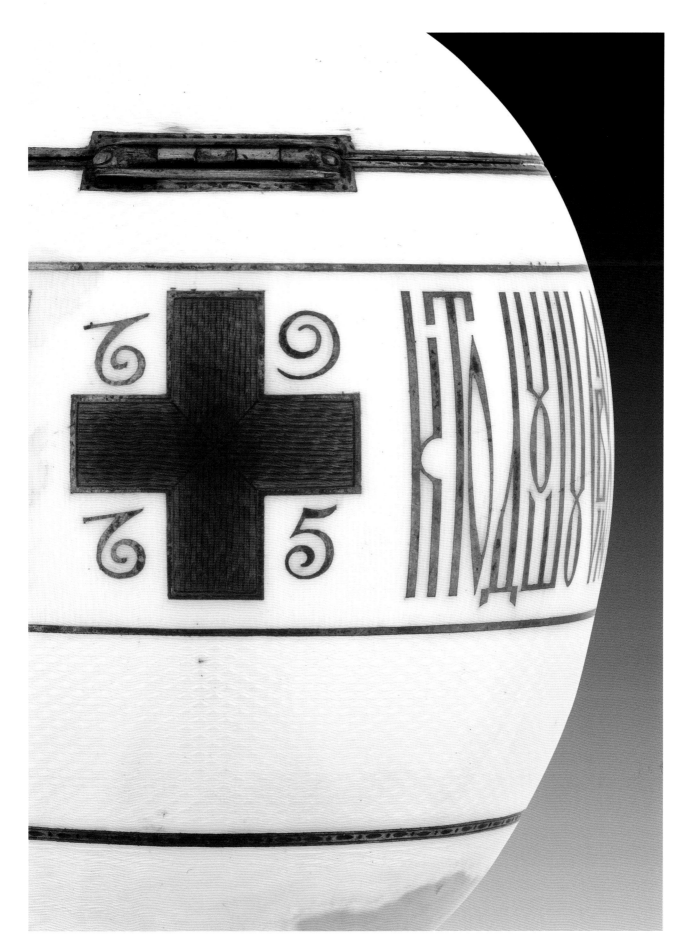

11

戒指盒

1899 年以前
法贝热公司
工匠大师：米哈伊尔·佩尔欣
黄金、红宝石、丝绸
高 34.925 毫米　直径 25.4 毫米
弗吉尼亚美术馆藏
莉莲·托马斯·普拉特遗赠

12

袖珍彩蛋吊坠

1899 年以前
法贝热公司
工匠大师：埃里克·科林
红玻璃、黄金
高 19.05 毫米　直径 15.875 毫米
弗吉尼亚美术馆藏
莉莲·托马斯·普拉特遗赠

13

袖珍彩蛋吊坠

1899 年以前
法贝热公司
黄金、红宝石、蓝宝石、祖母绿
高 15.875 毫米　直径 12.7 毫米
弗吉尼亚美术馆藏
莉莲·托马斯·普拉特遗赠

11

Ring Box

Before 1899
Fabergé firm
Workmaster: Mikhail Perkhin
Gold, ruby, silk
1 3/8" H × 1" Dia.
Virginia Museum of Fine Arts
Bequest of Lillian Thomas Pratt

12

Miniature Easter Egg Pendant

Before 1899
Fabergé firm
Workmaster: Erik Kollin
Purpurine, gold
3/4" H × 5/8" Dia.
Virginia Museum of Fine Arts
Bequest of Lillian Thomas Pratt

13

Miniature Easter Egg Pendant

Before 1899
Fabergé firm
Gold, rubies, sapphires, emerald
5/8" H × 1/2" Dia.
Virginia Museum of Fine Arts
Bequest of Lillian Thomas Pratt

057

14	15	16

袖珍彩蛋吊坠

1899 年以前
法贝热公司
水晶、黄金、红宝石、祖母绿
高 19.05 毫米　直径 15.875 毫米
弗吉尼亚美术馆藏
莉莲·托马斯·普拉特遗赠

袖珍彩蛋吊坠

1899 年以前
法贝热公司
黄金、银镀金、钻石、红宝石
高 28.575 毫米　直径 15.875 毫米
弗吉尼亚美术馆藏
莉莲·托马斯·普拉特遗赠

袖珍彩蛋吊坠

1899 年以前
法贝热公司
黄金、珐琅、月长石
高 19.05 毫米　直径 15.875 毫米
弗吉尼亚美术馆藏
小欧内斯特·希尔曼家族赠送

14

Miniature Easter Egg Pendant

Before 1899
Fabergé firm
Rock crystal, gold, rubies, emeralds
3/4" H × 5/8" Dia.
Virginia Museum of Fine Arts
Bequest of Lillian Thomas Pratt

15

Miniature Easter Egg Pendant

Before 1899
Fabergé firm
Gold, silver gilt, diamonds, ruby
1 1/8" H × 5/8" Dia.
Virginia Museum of Fine Arts
Bequest of Lillian Thomas Pratt

16

Miniature Easter Egg Pendant

Before 1899
Fabergé firm
Gold, enamel, moonstone
3/4" H × 5/8" Dia.
Virginia Museum of Fine Arts
Gift of the Estate of Ernest Hillman Jr.

17	18	19

袖珍彩蛋吊坠

1899 ~ 1908 年
法贝热公司
工匠大师：奥古斯特·霍姆斯特姆（？）
玛瑙、黄金、红宝石
高 15.875 毫米　直径 19.05 毫米
弗吉尼亚美术馆藏
莉莲·托马斯·普拉特遗赠

17 Miniature Easter Egg Pendant

1899–1908
Fabergé firm
Workmaster: August Holmström (?)
Agate, gold, rubies
5/8" H × 3/4" Dia.
Virginia Museum of Fine Arts
Bequest of Lillian Thomas Pratt

袖珍彩蛋吊坠

1899 ~ 1908 年
法贝热公司
玛瑙、黄金、银、珐琅、钻石
高 15.875 毫米　直径 12.7 毫米
弗吉尼亚美术馆藏
莉莲·托马斯·普拉特遗赠

18 Miniature Easter Egg Pendant

1899–1908
Fabergé firm
Agate, gold, silver, enamel, diamonds
5/8" H × 1/2" Dia.
Virginia Museum of Fine Arts
Bequest of Lillian Thomas Pratt

袖珍彩蛋吊坠

1899 ~ 1908 年
法贝热公司
工匠大师：亨瑞克·魏格斯特姆
黄金、珐琅
高 15.875 毫米　直径 12.7 毫米
弗吉尼亚美术馆藏
莉莲·托马斯·普拉特遗赠

19 Miniature Easter Egg Pendant

1899–1908
Fabergé firm
Workmaster: Henrik Wigström
Gold, enamel
5/8" H × 1/2" Dia.
Virginia Museum of Fine Arts
Bequest of Lillian Thomas Pratt

20

袖珍彩蛋吊坠
（带有皇后亚历山德拉·费奥多罗芙娜的花押）

约 1900 年
俄罗斯
工匠大师：A.A.（可能是安德烈·阿德勒）
黄金、水晶
高 15.875 毫米　直径 9.525 毫米
弗吉尼亚美术馆藏
莉莲·托马斯·普拉特遗赠

20

Miniature Easter
Egg Pendant
with Cypher for Tsaritsa
Alexandra Feodorovna

ca. 1900
Russian
Workmaster: A.A. (possibly Andrei K. Adler)
Gold, rock crystal
5/8" H × 3/8" Dia.
Virginia Museum of Fine Arts
Bequest of Lillian Thomas Pratt

21

袖珍彩蛋吊坠

约 1900 年
法贝热公司
黄金、珐琅
高 15.875 毫米　直径 12.7 毫米
弗吉尼亚美术馆藏
莉莲·托马斯·普拉特遗赠

21

Miniature Easter
Egg Pendant

ca. 1900
Fabergé firm
Gold, enamel
5/8" H × 1/2" Dia.
Virginia Museum of Fine Arts
Bequest of Lillian Thomas Pratt

22

袖珍彩蛋吊坠

约 1900 年
法贝热公司
黄金、珐琅
高 19.05 毫米　直径 15.875 毫米
弗吉尼亚美术馆藏
莉莲·托马斯·普拉特遗赠

22

Miniature Easter
Egg Pendant

ca. 1900
Fabergé firm
Gold, enamel
3/4" H × 5/8" Dia.
Virginia Museum of Fine Arts
Bequest of Lillian Thomas Pratt

23	24	25
袖珍彩蛋吊坠	**袖珍彩蛋吊坠**	**袖珍彩蛋吊坠**
约 1900 年 法贝热公司 黄金、珐琅、钻石、蓝宝石 高 19.05 毫米　直径 15.875 毫米 弗吉尼亚美术馆藏 莉莲·托马斯·普拉特遗赠	约 1900 年 法贝热公司 黄金、珐琅 高 19.05 毫米　直径 15.875 毫米 弗吉尼亚美术馆藏 莉莲·托马斯·普拉特遗赠	约 1900 年 法贝热公司 工匠大师：费奥多尔·阿法纳西耶夫 蔷薇辉石、黄金 高 19.05 毫米　直径 15.875 毫米 弗吉尼亚美术馆藏 莉莲·托马斯·普拉特遗赠
23	24	25
Miniature Easter Egg Pendant	**Miniature Easter Egg Pendant**	**Miniature Easter Egg Pendant**
ca. 1900 Fabergé firm Gold, enamel, diamonds, sapphires 3/4" H × 5/8" Dia. Virginia Museum of Fine Arts Bequest of Lillian Thomas Pratt	ca. 1900 Fabergé firm Gold, enamel 3/4" H × 5/8" Dia. Virginia Museum of Fine Arts Bequest of Lillian Thomas Pratt	ca. 1900 Fabergé firm Workmaster: Fedor Afanas'ev Rhodonite, gold 3/4" H × 5/8" Dia. Virginia Museum of Fine Arts Bequest of Lillian Thomas Pratt

26

袖珍彩蛋吊坠

约 1900 年
法贝热公司
欧泊、黄金
高 15.875 毫米　直径 12.7 毫米
弗吉尼亚美术馆藏
莉莲·托马斯·普拉特遗赠

27

袖珍彩蛋吊坠

约 1900 年
法贝热公司
黄水晶、黄金、银、珐琅、钻石
高 15.875 毫米　直径 9.525 毫米
弗吉尼亚美术馆藏
莉莲·托马斯·普拉特遗赠

28

袖珍彩蛋吊坠

约 1900 年
法贝热公司
黄金、珐琅
高 15.875 毫米　直径 9.525 毫米
弗吉尼亚美术馆藏
莉莲·托马斯·普拉特遗赠

26

Miniature Easter Egg Pendant

ca. 1900
Fabergé firm
Opal, gold
5/8" H × 1/2" Dia.
Virginia Museum of Fine Arts
Bequest of Lillian Thomas Pratt

27

Miniature Easter Egg Pendant

ca. 1900
Fabergé firm
Citrine, gold, silver, enamel, diamonds
5/8" H × 3/8" Dia.
Virginia Museum of Fine Arts
Bequest of Lillian Thomas Pratt

28

Miniature Easter Egg Pendant

ca.1900
Fabergé firm
Gold, enamel
5/8" H × 3/8" Dia.
Virginia Museum of Fine Arts
Bequest of Lillian Thomas Pratt

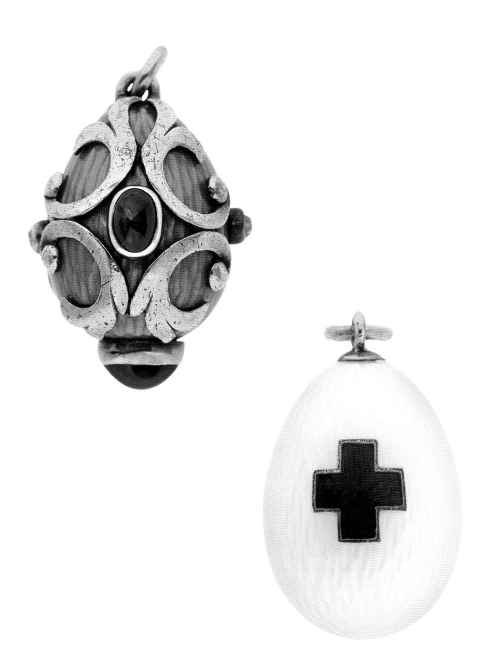

29 **袖珍彩蛋吊坠** 约 1900 年 法贝热公司 黄金、珐琅、银、珍珠、钻石 高 19.05 毫米　直径 15.875 毫米 弗吉尼亚美术馆藏 莉莲·托马斯·普拉特遗赠	30 **袖珍彩蛋吊坠** 约 1900 年 法贝热公司 黄金、银、珐琅、钻石、祖母绿 高 22.225 毫米　直径 14.288 毫米 弗吉尼亚美术馆藏 莉莲·托马斯·普拉特遗赠	31 **袖珍彩蛋吊坠** 约 1900 年 法贝热公司 工匠大师：米哈伊尔·佩尔欣 鲍文玉（硬绿蛇纹石）、黄金、红宝石、钻石 高 19.05 毫米　直径 15.875 毫米 弗吉尼亚美术馆藏 莉莲·托马斯·普拉特遗赠

29
Miniature Easter Egg Pendant

ca. 1900
Fabergé firm
Gold, enamel, silver, pearls, diamonds
3/4" H × 5/8" Dia.
Virginia Museum of Fine Arts
Bequest of Lillian Thomas Pratt

30
Miniature Easter Egg Pendant

ca. 1900
Fabergé firm
Gold, silver, enamel, diamonds, emerald
7/8" H × 9/16" Dia.
Virginia Museum of Fine Arts
Bequest of Lillian Thomas Pratt

31
Miniature Easter Egg Pendant

ca. 1900
Fabergé firm
Workmaster: Mikhail Perkhin
Bowenite, gold, rubies, diamonds
3/4" H × 5/8" Dia.
Virginia Museum of Fine Arts
Bequest of Lillian Thomas Pratt

| 32 | 33 | 34 |

袖珍彩蛋吊坠

约 1900 年
法贝热公司
工匠大师：阿尔佛雷德·蒂勒曼
黄金、珐琅、红宝石
高 19.05 毫米　直径 15.875 毫米
弗吉尼亚美术馆藏
莉莲·托马斯·普拉特遗赠

袖珍彩蛋吊坠

约 1900 年
法贝热公司
工匠大师：费奥多尔·阿法纳西耶夫
黄金、珐琅
高 19.05 毫米　直径 15.875 毫米
弗吉尼亚美术馆藏
莉莲·托马斯·普拉特遗赠

袖珍彩蛋吊坠

约 1900 年
法贝热公司
工匠大师：米哈伊尔·佩尔欣
玉髓、黄金、白金、钻石
高 31.75 毫米　直径 22.225 毫米
弗吉尼亚美术馆藏
莉莲·托马斯·普拉特遗赠

32 Miniature Easter Egg Pendant

ca. 1900
Fabergé firm
Workmaster: Alfred Thielemann
Gold, enamel, rubies
3/4" H × 5/8" Dia.
Virginia Museum of Fine Arts
Bequest of Lillian Thomas Pratt

33 Miniature Easter Egg Pendant

ca. 1900
Fabergé firm
Workmaster: Fedor Afanas'ev
Gold, enamel
3/4" H × 5/8" Dia.
Virginia Museum of Fine Arts
Bequest of Lillian Thomas Pratt

34 Miniature Easter Egg Pendant

ca. 1900
Fabergé firm
Workmaster: Mikhail Perkhin
Chalcedony, gold, white gold, diamonds
1 1/4" H × 7/8" Dia.
Virginia Museum of Fine Arts
Bequest of Lillian Thomas Pratt

美轮美奂的装饰艺术
Majestic Decorative Arts

彼得·卡尔·法贝热1846年出生在圣彼得堡。青年时期的法贝热曾游历欧洲各国，到1872年继承家族珠宝作坊时，已经具备了丰富的商业知识和非凡的艺术眼光。当时俄国权贵们在珠宝首饰方面的品位不高，珠宝价值主要取决于材料的尺寸和重量。在这种华而不实、缺乏艺术修养的风气中，法贝热发动了一场艺术风格的革命。他将重点由珠宝的克拉数转移到对艺术创造性和工艺水平的追求上，不断拓展选材和创作主题的范围，打造出一系列风格清新、独具创意的艺术珍品。

Peter Karl Fabergé was born in Saint Petersburg, Russia, in 1846. He made a Grand Tour of Europe while a young man, and when taking over the family jewelry business of House of Fabergé in 1872, he had acquired a wealth of business knowledge and an extraordinary vision of art. The Russian high society lacked taste in jewelry then, and the value of jewels hinged dominantly on the size and weight of material used for jewels. Recognizing such ostentation and lack of artistry, Fabergé launched a revolution in the style of art. He shifted his focus from the carats of jewels to pursuing art creativity and craftsmanship, expanded the range of materials and subjects for his creation, and eventually created a series of art objects of fresh styles and unique ideas.

35

钻石别针

19 世纪
俄罗斯
黄金、钻石
高 25.4 毫米　宽 19.05 毫米　厚 3.175 毫米
弗吉尼亚美术馆藏
莉莲·托马斯·普拉特遗赠

35

Diamond Pin

19th Century
Russian
Gold, diamonds
1" H × 3/4" W × 1/8" D
Virginia Museum of Fine Arts
Bequest of Lillian Thomas Pratt

36

皇冠胸针

1890 ~ 1910 年
俄罗斯
银镀金、蓝宝石、红宝石、钻石
高 61.913 毫米　宽 77.788 毫米　厚 19.05 毫米
弗吉尼亚美术馆藏
莉莲·托马斯·普拉特遗赠

36

Crown Brooch

1890–1910
Russian
Silver gilt, sapphires, rubies, diamonds
2 7/16" H × 3 1/16" W × 3/4" D
Virginia Museum of Fine Arts
Bequest of Lillian Thomas Pratt

37

皇家钻石胸针

1890 ~ 1910 年
俄罗斯
黄金、银、珐琅、钻石
高 28.575 毫米　宽 38.1 毫米　厚 9.525 毫米
弗吉尼亚美术馆藏
莉莲·托马斯·普拉特遗赠

37

Imperial Diamond Brooch

1890 – 1910
Russian
Gold, silver, enamel, diamonds
1 1/8" H × 1 1/2" W × 3/8" D
Virginia Museum of Fine Arts
Bequest of Lillian Thomas Pratt

38

十周年纪念胸针

1899 ~ 1908 年
法贝热公司
工匠大师：费奥多尔·阿法纳西耶夫
黄金、银、珐琅、钻石
高 28.575 毫米　宽 34.925 毫米　厚 9.525 毫米
弗吉尼亚美术馆藏
莉莲·托马斯·普拉特遗赠

　　这件八角形胸针的中央是一个钻石镶嵌的罗马数字"X"，周围环绕着一圈月桂花环，胸针边缘镶嵌着十颗明亮的切割钻石。这件胸针是为了纪念某个十周年纪念日，或许是为某位皇室成员制作。1904 年是沙皇和皇后结婚十周年，也是令这对皇室夫妇悲喜交加的一年。1 月，日本驱逐舰突然对三艘俄罗斯战舰发动攻击，导致灾难性的日俄战争爆发。7 月，他们的儿子暨期待已久的皇位继承人——阿列克谢·尼古拉耶维奇诞生，这让他们能够暂时把烦恼搁置一旁，以庆祝这一幸事。

38

Tenth-Anniversary Brooch

1899–1908
Fabergé firm
Workmaster: Fedor Afanas'ev
Gold, silver, enamel, diamonds
1 1/8" H × 1 3/8" W × 3/8" D
Virginia Museum of Fine Arts
Bequest of Lillian Thomas Pratt

　　This octagonal brooch has a diamond-set Roman numeral X within a laurel wreath and its borders are set with ten brilliant cut diamonds. It was created to commemorate a tenth anniversary, perhaps that of a member of the imperial court. The year 1904, during which the tsar and tsaritsa celebrated their tenth wedding anniversary, brought both pain and joy into the life of the royal couple. The disastrous Russo-Japanese war broke out in January, triggered by a surprise attack on three Russian battleships by Japanese destroyers. In July, however, their troubles were briefly set aside to celebrate the birth of their son—and long awaited heir—Aleksei Nikolaevich.

39

圣甲虫胸针

约 1900 年

法贝热公司

石榴石、黄金、钻石、红宝石、珐琅、银

高 28.575 毫米　宽 38.1 毫米　厚 19.05 毫米

弗吉尼亚美术馆藏

莉莲·托马斯·普拉特遗赠

40

野玫瑰胸针

约 1900 年

法贝热公司

黄金、珐琅

直径 41.275 毫米　厚 12.7 毫米

弗吉尼亚美术馆藏

莉莲·托马斯·普拉特遗赠

39

Scarab Brooch

ca. 1900

Fabergé firm

Garnet, gold, diamonds, rubies, enamel, silver

1 1/8" H × 1 1/2" W × 3/4" D

Virginia Museum of Fine Arts

Bequest of Lillian Thomas Pratt

40

Wild Rose Brooch

ca. 1900

Fabergé firm

Gold, enamel

1 5/8" Dia. × 1/2" D

Virginia Museum of Fine Arts

Bequest of Lillian Thomas Pratt

41

印章

1899 年以前
法贝热公司
工匠大师：米哈伊尔·佩尔欣
鲍文玉（硬绿蛇纹石）、黄金、银镀金、珍珠、玉髓
高 76.2 毫米　直径 31.75 毫米
弗吉尼亚美术馆藏
莉莲·托马斯·普拉特遗赠

42

印章

约 1900 年
法贝热公司
工匠大师：埃里克·科林
东陵玉、黄金、银
高 79.375 毫米　宽 50.8 毫米　厚 44.45 毫米
弗吉尼亚美术馆藏
莉莲·托马斯·普拉特遗赠

41

Seal

Before 1899
Fabergé firm
Workmaster: Mikhail Perkhin
Bowenite, gold, silver gilt, pearls, chalcedony
3" H × 1 1/4" Dia.
Virginia Museum of Fine Arts
Bequest of Lillian Thomas Pratt

42

Seal

ca. 1900
Fabergé firm
Workmaster: Erik Kollin
Aventurine, gold, silver
3 1/8" H × 2" W × 1 3/4" D
Virginia Museum of Fine Arts
Bequest of Lillian Thomas Pratt

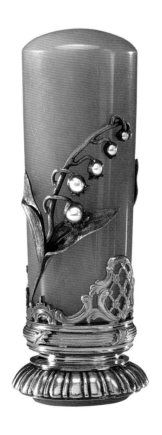

43

双幅相框

1899 年以前
法贝热公司
工匠大师：米哈伊尔·佩尔欣
鲍文玉（硬绿蛇纹石）、黄金、水晶
高 57.15 毫米　宽 88.9 毫米　厚 38.1 毫米
弗吉尼亚美术馆藏
莉莲·托马斯·普拉特遗赠

43

Double Frame

Before 1899
Fabergé firm
Workmaster: Mikhail Perkhin
Bowenite, gold, rock crystal
2 1/4" H × 3 1/2" W × 1 1/2" D
Virginia Museum of Fine Arts
Bequest of Lillian Thomas Pratt

44

双幅相框

1899～1908 年
法贝热公司
工匠大师：亨瑞克·魏格斯特姆
鲍文玉（硬绿蛇纹石）、银镀金、珐琅、猛犸象牙、玻璃
高 76.2 毫米　宽 114.3 毫米　厚 15.875 毫米
弗吉尼亚美术馆藏
莉莲·托马斯·普拉特遗赠

44

Double Frame

1899–1908
Fabergé firm
Workmaster: Henrik Wigström
Bowenite, silver gilt, enamel, Mammoth ivory, glass
3" H × 4 1/2" W × 5/8" D
Virginia Museum of Fine Arts
Bequest of Lillian Thomas Pratt

45

双幅相框

1899 ~ 1908 年
法贝热公司
工匠大师：亚尔马·阿姆菲尔特
软玉、黄金、银、银镀金、珍珠、
红宝石、玻璃、螺钿
高 88.9 毫米　宽 123.825 毫米　厚 15.875 毫米
弗吉尼亚美术馆藏
莉莲·托马斯·普拉特遗赠

45

Double Frame

1899–1908
Fabergé firm
Workmaster: Hjalmar Armfelt
Nephrite, gold, silver, silver gilt, pearls,
rubies, glass, mother-of-pearl
3 1/2" H × 4 7/8" W × 5/8" D
Virginia Museum of Fine Arts
Bequest of Lillian Thomas Pratt

46

双幅相框

19 ~ 20 世纪，1947 年以前
未知
法贝热仿制品
银镀金、珐琅、红宝石、钻石、猛犸象牙、玻璃
高 107.95 毫米　宽 152.4 毫米
弗吉尼亚美术馆藏
莉莲·托马斯·普拉特遗赠

　　这件玑镂珐琅相框饰有一只双头鹰，上方镶有三枚红宝石，下方则是玫瑰花结和月桂垂花饰。相框顶部环绕着两个鹰头狮身怪兽。1947 年，莉莲·托马斯·普拉特将这组俄罗斯艺术品收藏系列捐赠给弗吉尼亚美术馆，其中包括部分法贝热赝品。这件相框的品质实在过于粗糙，无法令人相信是由圣彼得堡领军珠宝行制作的。况且，法贝热作品上通常标有工匠大师的姓名缩写（如卡尔·布朗克的 CB，或亚历山大·特雷登的 AT），而这件作品上并无此类印记。双头鹰是皇室物品的象征，但通常只用于香烟盒或珠宝首饰，从不会用于相框。

46

Double Frame

19th–20th Century, before 1947
Unknown
Fabergé forgery
Silver gilt, enamel, rubies, diamond, Mammoth ivory, glass
4 1/4" H × 6" W
Virginia Museum of Fine Arts
Bequest of Lillian Thomas Pratt

　　This guilloché enamel frame is decorated with a double-headed eagle set with three rubies above, and three laurel swags suspended from rosettes below. It is surmounted by two griffins. The Russian-art collection bequeathed in 1947 by Lillian Thomas Pratt to the Virginia Museum of Fine Arts includes its share of Fauxbergé. The quality of this frame is too poor for one of the leading workshops in St. Petersburg. It lacks the usual workmaster's initials (CB for Carl Blank or AT for Alexander Treyden). The double-headed eagle also denotes an imperial presentation, which normally only appears on cigarette cases or on items of jewelry, but never on frames.

47

三幅相框

1899～1908年
法贝热公司
黄金、珐琅、钻石、玻璃、猛犸象牙
高 57.15 毫米　宽 57.15 毫米　厚 12.7 毫米
弗吉尼亚美术馆藏
莉莲·托马斯·普拉特遗赠

47

Triple Frame

1899–1908
Fabergé firm
Gold, enamel, diamonds, glass, Mammoth ivory
2 1/4" H × 2 1/4" W × 1/2" D
Virginia Museum of Fine Arts
Bequest of Lillian Thomas Pratt

48

三联相框

1899～1908年
法贝热公司
工匠大师：亚尔马·阿姆菲尔特
银、银镀金、珐琅、木、玻璃
高114.3毫米　宽260.35毫米　厚12.7毫米
弗吉尼亚美术馆藏
莉莲·托马斯·普拉特遗赠

48

Triptych Frame

1899–1908
Fabergé firm
Workmaster: Hjalmar Armfelt
Silver, silver gilt, enamel, wood, glass
4 1/2" H × 10 1/4" W × 1/2" D
Virginia Museum of Fine Arts
Bequest of Lillian Thomas Pratt

1899年以前
法贝热公司
工匠大师：米哈伊尔·佩尔欣
黄金、银、珐琅、珍珠、玻璃、赛璐珞（现代）
高 80.963 毫米　宽 73.025 毫米　厚 12.7 毫米
弗吉尼亚美术馆藏
莉莲·托马斯·普拉特遗赠

　　米哈伊尔·佩尔欣工坊制作的作品有鼻烟壶、香烟盒、歌剧眼镜、扇子、座钟、纸牌盒、阳伞和手杖柄以及无数小型珍品，大多数为华丽的珐琅彩，并采用洛可可或路易十五时期的风格装饰。为迎合当时新兴的摄影潮流，佩尔欣也制作了许多相框，这张嵌有沙皇尼古拉二世二女儿塔季扬娜女大公相片的相框就是其中之一。这件星形相框由沙皇和皇后于1896年共同购买，沙皇夫妇被流放西伯利亚时也随身携带着这只相框。

Before 1899
Fabergé firm
Workmaster: Mikhail Perkhin
Gold, silver, enamel, pearls, glass, celluloid (modern)
3 3/16" H × 2 7/8" W × 1/2" D
Virginia Museum of Fine Arts
Bequest of Lillian Thomas Pratt

　　Objects created by Mikhail Perkhin's workshop, most of them in lush enamel colors and decorated in the rococo or Louis-XV taste, included snuffboxes, cigarette cases, opera glasses, fans, desk clocks, card cases, parasol and cane handles, and countless small *objets de vertu* (objects of virtue). In response to the new craze for photographs, Perkhin also produced numerous frames, such as this one containing an original photograph of Grand Duchess Tatiana, the second daughter of Tsar Nicholas II. The star-shaped frame was acquired jointly by the tsar and tsaritsa in 1896 and accompanied them into exile in Siberia.

50

心形相框

约 1900 年
法贝热公司
工匠大师：亨瑞克·魏格斯特姆
软玉、黄金、水晶、绸缎
高 85.725 毫米　宽 73.025 毫米　厚 28.575 毫米
弗吉尼亚美术馆藏
莉莲·托马斯·普拉特遗赠

50

Heart Frame

ca. 1900
Fabergé firm
Workmaster: Henrik Wigström
Nephrite, gold, rock crystal, satin
3 3/8" H × 2 7/8" W × 1 1/8" D
Virginia Museum of Fine Arts
Bequest of Lillian Thomas Pratt

51

皇家银相框

1899～1908 年
法贝热公司
银、玛瑙、蓝宝石、玉髓、木、玻璃
高 444.5 毫米　宽 295.275 毫米　厚 34.925 毫米
弗吉尼亚美术馆藏
佛利斯特·马尔斯、约翰·马尔斯和杰奎琳·马尔斯捐赠，以纪念其母奥黛丽·马尔斯与其父老佛利斯特·马尔斯

　　这个银质相框的造型仿佛一扇门，门框四周装饰着抽象的叶子和玫瑰花饰。相框上装饰着凸面宝石，有玛瑙、蓝宝石和玉髓。相框顶部栖息着一只采用透雕工艺雕刻的加冕双头鹰，鹰身刻有尼古拉二世姓名的交织字母。这是一个将传统俄罗斯装饰艺术与皇室象征完美结合的典范。

51

Imperial Silver Frame

1899–1908
Fabergé firm
Silver, agates, sapphires, chalcedonies, wood, glass
17 1/2" H × 11 5/8" W × 1 3/8" D
Virginia Museum of Fine Arts
Gift of Forrest E. Mars, John F. Mars, and Jacqueline B. Mars in honor of their mother and father, Audrey M. Mars and Forrest E. Mars Sr.

　　This silver frame is shaped like a doorway decorated with stylized foliage and rosettes. It is ornamented with cabochon stones, including agates, sapphires, and chalcedonies. The frame is surmounted by a crowned openwork double-headed eagle with the monogram of Tsar Nicholas II. This work of art is a representative example of Old-Russian ornament combined with imperial symbols.

52

十周年纪念相框

1899～1908 年
法贝热公司
工匠大师：亨瑞克·魏格斯特姆
银、银镀金、珐琅、钻石、玻璃、猛犸象牙
高 88.9 毫米　宽 88.9 毫米　厚 12.7 毫米
弗吉尼亚美术馆藏
莉莲·托马斯·普拉特遗赠

52

Tenth-Anniversary Frame

1899–1908
Fabergé firm
Workmaster: Henrik Wigström
Silver, silver gilt, enamel, diamonds, glass, Mammoth ivory
3 1/2" H × 3 1/2" W × 1/2" D
Virginia Museum of Fine Arts
Bequest of Lillian Thomas Pratt

53

圣安德烈十字相框

1899～1908 年
法贝热公司
工匠大师：安蒂·内瓦莱南
银镀金、珐琅、木、丝绸、玻璃
高 149.225 毫米　宽 212.725 毫米　厚 15.875 毫米
弗吉尼亚美术馆藏
莉莲·托马斯·普拉特遗赠

此相框中的蓝色纹章源自苏格兰圣安德烈的"X"形十字架，它也是彼得大帝于1698年设立的圣安德烈勋章的组成部分。沙皇尼古拉二世是俄罗斯最高荣誉——圣安德烈勋章的授予者。里面镶嵌的相片拍摄于1916年，是从一张明信片上剪下来的，相片中为打扮成慈悲修女的奥尔加女大公和塔季扬娜女大公，以及她们的妹妹玛利亚和阿纳斯塔西亚。

53

Cross of St. Andrew Frame

1899–1908
Fabergé firm
Workmaster: Antti Nevalainen
Silver gilt, enamel, wood, silk, glass
5 7/8" H × 8 3/8" W × 5/8" D
Virginia Museum of Fine Arts
Bequest of Lillian Thomas Pratt

The blue heraldic device featured in this frame is inspired by the Scottish version of St. Andrew's X-shaped cross. It formed part of the emblem of the Order of St. Andrew founded by Tsar Peter the Great in 1698. As tsar, Nicholas II was Grand Master of this, Russia's highest order. The 1916 photograph of Grand Duchesses Olga and Tatiana as Sisters of Mercy and their younger sisters Maria and Anastasia are cut from a postcard.

54

皇家柱形相框

1908 年
法贝热公司
工匠大师：亨瑞克·魏格斯特姆
黄金、银、钻石、猛犸象牙、水彩
相框：高 152.4 毫米　长 55.6 毫米　宽 55.6 毫米
微型画：高 31.75 毫米　宽 25.4 毫米
弗吉尼亚美术馆藏
莉莲·托马斯·普拉特遗赠

　　这件罕见的柱形相框是沙皇尼古拉二世专为陆军元帅德米特里·阿列克谢耶维奇·米柳京伯爵定制，以纪念他从军 75 周年。一身戎装的米柳京伯爵于 1861 年至 1881 年期间担任陆军大臣。迄今已知的柱形相框仅有五件，全部都由法贝热的亨瑞克·魏格斯特姆工坊制作。

54

Imperial Column Portrait Frame

1908
Fabergé firm
Workmaster: Henrik Wigström
Gold, silver, diamonds, Mammoth ivory, watercolor
Frame: 6" H × 2 3/16" L × 2 3/16" W
Miniature: 1 1/4" H × 1" W
Virginia Museum of Fine Arts
Bequest of Lillian Thomas Pratt

　　This rare column portrait frame was specially commissioned by Tsar Nicholas II for Field Marshal-General Count Dmitrii Alekseevich Miliutin for his seventy-fifth anniversary as an officer. The highly decorated Count Miliutin was minister of war between 1861 and 1881. There are only five known examples of column portrait frames, all of them made by Fabergé in the workshop of Henrik Wigström.

55

柱形相框

1908～1917年
法贝热公司
工匠大师：亨瑞克·魏格斯特姆
鲍文玉（硬绿蛇纹石）、黄金、水晶、红宝石、珍珠
高 111.125 毫米　直径 41.275 毫米
弗吉尼亚美术馆藏
莉莲·托马斯·普拉特遗赠

55

Column Frame

1908–1917
Fabergé firm
Workmaster: Henrik Wigström
Bowenite, gold, rock crystal, rubies, pearl
4 3/8" H × 1 5/8" Dia.
Virginia Museum of Fine Arts
Bequest of Lillian Thomas Pratt

56

可翻转相框

1908～1917 年
法贝热公司
工匠大师：亚尔马·阿姆菲尔特
银镀金、珐琅、玻璃
高 104.775 毫米　宽 76.2 毫米　厚 9.525 毫米
弗吉尼亚美术馆藏
莉莲·托马斯·普拉特遗赠

56

Reversible Frame

1908–1917
Fabergé firm
Workmaster: Hjalmar Armfelt
Silver gilt, enamel, glass
4 1/8" H × 3" W × 3/8" D
Virginia Museum of Fine Arts
Bequest of Lillian Thomas Pratt

57

相框

1895 ~ 1916 年
法贝热公司
工匠大师：亚尔马·阿姆菲尔特
青金石、黄金、银、
珍珠、钻石、玻璃、螺钿
高 66.675 毫米　宽 60.325 毫米　厚 12.7 毫米
弗吉尼亚美术馆藏
莉莲·托马斯·普拉特遗赠

57

Frame

1895–1916
Fabergé firm
Workmaster: Hjalmar Armfelt
Lapis lazuli, gold, silver, pearls,
diamonds, glass, mother-of-pearl
2 5/8" H × 2 3/8" W × 1/2" D
Virginia Museum of Fine Arts
Bequest of Lillian Thomas Pratt

58

相框

1899 年以前
法贝热公司
工匠大师：米哈伊尔·佩尔欣
银镀金、珐琅、珍珠、水晶、猛犸象牙
高 114.3 毫米　宽 87.313 毫米　厚 12.7 毫米
弗吉尼亚美术馆藏
莉莲·托马斯·普拉特遗赠

58

Frame

Before 1899
Fabergé firm
Workmaster: Mikhail Perkhin
Silver gilt, enamel, pearls, rock crystal, Mammoth ivory
4 1/2" H × 3 7/16" W × 1/2" D
Virginia Museum of Fine Arts
Bequest of Lillian Thomas Pratt

59

相框

1899 年以前

法贝热公司

工匠大师：米哈伊尔·佩尔欣

银镀金、珐琅、珍珠、猛犸象牙、玻璃

高 92.075 毫米　宽 66.675 毫米　厚 9.525 毫米

弗吉尼亚美术馆藏

莉莲·托马斯·普拉特遗赠

59

Frame

Before 1899

Fabergé firm

Workmaster: Mikhail Perkhin

Silver gilt, enamel, pearls, Mammoth ivory, glass

3 5/8" H × 2 5/8" W × 3/8" D

Virginia Museum of Fine Arts

Bequest of Lillian Thomas Pratt

60

相框

1899 年以前

法贝热公司

工匠大师：米哈伊尔·佩尔欣

黄金、银镀金、黄铜、猛犸象牙、玻璃

照片：阿娜斯塔西娅女大公

高 82.55 毫米　宽 69.85 毫米　厚 12.7 毫米

弗吉尼亚美术馆藏

莉莲·托马斯·普拉特遗赠

60

Frame

Before 1899

Fabergé firm

Workmaster: Mikhail Perkhin

Gold, silver gilt, brass, Mammoth ivory, glass

Photo: Grand Duchess Ansatasia

3 1/4" H × 2 3/4" W × 1/2" D

Virginia Museum of Fine Arts

Bequest of Lillian Thomas Pratt

61

相框

1899 年以前
法贝热公司
工匠大师：米哈伊尔·佩尔欣
黄金、银镀金、珐琅、猛犸象牙、玻璃
高 85.725 毫米　宽 73.025 毫米　厚 9.525 毫米
弗吉尼亚美术馆藏
莉莲·托马斯·普拉特遗赠

61

Frame

Before 1899
Fabergé firm
Workmaster: Mikhail Perkhin
Gold, silver gilt, enamel, Mammoth ivory, glass
3 3/8" H × 2 7/8" W × 3/8" D
Virginia Museum of Fine Arts
Bequest of Lillian Thomas Pratt

62

相框

1899 年以前
法贝热公司
工匠大师：米哈伊尔·佩尔欣
东陵玉、银镀金、玻璃、织物
高 82.55 毫米　宽 60.325 毫米　厚 12.7 毫米
弗吉尼亚美术馆藏
莉莲·托马斯·普拉特遗赠

62

Frame

Before 1899
Fabergé firm
Workmaster: Mikhail Perkhin
Aventurine, silver gilt, glass, fabric
3 1/4" H × 2 3/8" W × 1/2" D
Virginia Museum of Fine Arts
Bequest of Lillian Thomas Pratt

63

相框

1899 年以前
俄罗斯
亚历山大·特雷顿
黄金、珐琅、猛犸象牙、玻璃
高 79.375 毫米　宽 66.675 毫米　厚 9.525 毫米
弗吉尼亚美术馆藏
莉莲·托马斯·普拉特遗赠

63

Frame

Before 1899
Russian
Alexander Treyden
Gold, enamel, Mammoth ivory, glass
3 1/8" H × 2 5/8" W × 3/8" D
Virginia Museum of Fine Arts
Bequest of Lillian Thomas Pratt

64

相框

1899 年以前
法贝热公司
工匠大师：米哈伊尔·佩尔欣
鲍文玉（硬绿蛇纹石）、黄金、玻璃、猛犸象牙
高 76.2 毫米　宽 76.2 毫米　厚 12.7 毫米
弗吉尼亚美术馆藏
莉莲·托马斯·普拉特遗赠

64

Frame

Before 1899
Fabergé firm
Workmaster: Mikhail Perkhin
Bowenite, gold, glass, Mammoth ivory
3" H × 3" W × 1/2" D
Virginia Museum of Fine Arts
Bequest of Lillian Thomas Pratt

65

相框

1899 年以前

法贝热公司

工匠大师：维科托·阿尔内

银镀金、珐琅、珍珠、玻璃、猛犸象牙

高 79.375 毫米　宽 88.9 毫米　厚 12.7 毫米

弗吉尼亚美术馆藏

莉莲·托马斯·普拉特遗赠

65

Frame

Before 1899

Fabergé firm

Workmaster: Viktor Aarne

Silver gilt, enamel, pearls, glass, Mammoth ivory

3 1/8" H × 3 1/2" W × 1/2" D

Virginia Museum of Fine Arts

Bequest of Lillian Thomas Pratt

66

相框

1899 年以前

法贝热公司

工匠大师：米哈伊尔·佩尔欣

黄金、珐琅、银、钻石、玻璃、猛犸象牙

照片：塔季扬娜女大公

高 73.025 毫米　宽 60.325 毫米　厚 15.875 毫米

弗吉尼亚美术馆藏

莉莲·托马斯·普拉特遗赠

66

Frame

Before 1899

Fabergé firm

Workmaster: Mikhail Perkhin

Gold, enamel, silver, diamonds, glass, Mammoth ivory

Photo: Grand Duchess Tatiana

2 7/8" H × 2 3/8" W × 5/8" D

Virginia Museum of Fine Arts

Bequest of Lillian Thomas Pratt

67

相框

1899 年以前
法贝热公司
工匠大师：米哈伊尔・佩尔欣
黄金、珐琅、珍珠、猛犸象牙、玻璃
高 85.725 毫米　宽 85.725 毫米
弗吉尼亚美术馆藏
莉莲・托马斯・普拉特遗赠

67

Frame

Before 1899
Fabergé firm
Workmaster: Mikhail Perkhin
Gold, enamel, pearls, Mammoth ivory, glass
3 3/8" H × 3 3/8" W
Virginia Museum of Fine Arts
Bequest of Lillian Thomas Pratt

68

相框

1899～1908年
法贝热公司
黄金、珐琅、钻石、猛犸象牙、玻璃
高139.7毫米　宽92.075毫米　厚12.7毫米
弗吉尼亚美术馆藏
莉莲·托马斯·普拉特遗赠

68

Frame

1899–1908
Fabergé firm
Gold, enamel, diamonds, Mammoth ivory, glass
5 1/2" H × 3 5/8" W × 1/2" D
Virginia Museum of Fine Arts
Bequest of Lillian Thomas Pratt

| 69 | 69 |

相框

1899～1908 年
法贝热公司
工匠大师：维克托·阿尔内
银镀金、珐琅、木、玻璃
高 142.875 毫米　宽 117.475 毫米　厚 15.875 毫米
弗吉尼亚美术馆藏
莉莲·托马斯·普拉特遗赠

Frame

1899–1908
Fabergé firm
Workmaster: Viktor Aarne
Silver gilt, enamel, wood, glass
5 5/8" H × 4 5/8" W × 5/8" D
Virginia Museum of Fine Arts
Bequest of Lillian Thomas Pratt

70

相框

1899～1908年
法贝热公司
工匠大师：米哈伊尔·佩尔欣
黄金、珐琅、猛犸象牙、玻璃
高 92.075 毫米　宽 66.675 毫米　厚 15.875 毫米
弗吉尼亚美术馆藏
莉莲·托马斯·普拉特遗赠

70

Frame

1899–1908
Fabergé firm
Workmaster: Mikhail Perkhin
Gold, enamel, Mammoth ivory, glass
3 5/8" H × 2 5/8" W × 5/8" D
Virginia Museum of Fine Arts
Bequest of Lillian Thomas Pratt

71

相框

1899～1908年
法贝热公司
工匠大师：亨瑞克·魏格斯特姆
软玉、黄金、银镀金、猛犸象牙、玻璃
高 139.7 毫米　宽 69.85 毫米
弗吉尼亚美术馆藏
莉莲·托马斯·普拉特遗赠

71

Frame

1899–1908
Fabergé firm
Workmaster: Henrik Wigström
Nephrite, gold, silver gilt, Mammoth ivory, glass
5 1/2" H × 2 3/4" W
Virginia Museum of Fine Arts
Bequest of Lillian Thomas Pratt

72

相框

1899～1908 年
法贝热公司
工匠大师：亨瑞克·魏格斯特姆
黄金、银镀金、珐琅、珍珠、玻璃、猛犸象牙
直径 82.55 毫米　厚 15.875 毫米
弗吉尼亚美术馆藏
莉莲·托马斯·普拉特遗赠

72

Frame

1899–1908
Fabergé firm
Workmaster: Henrik Wigström
Gold, silver gilt, enamel, pearls, glass, Mammoth ivory
3 1/4" Dia. × 5/8" D
Virginia Museum of Fine Arts
Bequest of Lillian Thomas Pratt

73

相框

1899～1908 年
法贝热公司
工匠大师：亨瑞克·魏格斯特姆
黄金、珐琅、银、珍珠、玻璃、猛犸象牙
直径 82.55 毫米　厚 12.7 毫米
弗吉尼亚美术馆藏
莉莲·托马斯·普拉特遗赠

73

Frame

1899–1908
Fabergé firm
Workmaster: Henrik Wigström
Gold, enamel, silver, pearls, glass, Mammoth ivory
3 1/4" Dia. × 1/2" D
Virginia Museum of Fine Arts
Bequest of Lillian Thomas Pratt

74

相框

1899 ~ 1908 年
法贝热公司
工匠大师：亨瑞克·魏格斯特姆
黄金、银镀金、珐琅、珍珠、玻璃、猛犸象牙
直径 76.2 毫米　厚 9.525 毫米
弗吉尼亚美术馆藏
莉莲·托马斯·普拉特遗赠

74

Frame

1899–1908
Fabergé firm
Workmaster: Henrik Wigström
Gold, silver gilt, enamel, pearls, glass, Mammoth ivory
3" Dia. × 3/8" D
Virginia Museum of Fine Arts
Bequest of Lillian Thomas Pratt

75

相框

1899 ~ 1908 年
法贝热公司
工匠大师：亨瑞克·魏格斯特姆
银镀金、黄金、珐琅、玻璃、猛犸象牙、水彩
高 53.975 毫米　宽 47.675 毫米　厚 9.525 毫米
弗吉尼亚美术馆藏
莉莲·托马斯·普拉特遗赠

75

Frame

1899–1908
Fabergé firm
Workmaster: Henrik Wigström
Silver gilt, gold, enamel, glass, Mammoth ivory, watercolor
2 1/8" H × 1 7/8" W × 3/8" D
Virginia Museum of Fine Arts
Bequest of Lillian Thomas Pratt

76

相框

1899～1908 年
法贝热公司
工匠大师：亨瑞克·魏格斯特姆
黄金、银、珐琅、红宝石、珍珠、玻璃、猛犸象牙
高 79.375 毫米　宽 57.15 毫米　厚 9.525 毫米
弗吉尼亚美术馆藏
莉莲·托马斯·普拉特遗赠

76

Frame

1899–1908
Fabergé firm
Workmaster: Henrik Wigström
Gold, silver, enamel, rubies, pearls, glass, Mammoth ivory
3 1/8" H × 2 1/4" W × 3/8" D
Virginia Museum of Fine Arts
Bequest of Lillian Thomas Pratt

77

相框

1899～1908 年
法贝热公司
工匠大师：安蒂·内瓦莱南
银镀金、珐琅、玻璃、木、黄金、猛犸象牙、水彩
高 95.25 毫米　宽 76.2 毫米　厚 12.7 毫米
弗吉尼亚美术馆藏
莉莲·托马斯·普拉特遗赠

77

Frame

1899–1908
Fabergé firm
Workmaster: Antti Nevalainen
Silver gilt, enamel, glass, wood, gold, Mammoth ivory, watercolor
3 3/4" H × 3" W × 1/2" D
Virginia Museum of Fine Arts
Bequest of Lillian Thomas Pratt

78

相框

1899～1908 年
法贝热公司
银、猛犸象牙、珐琅、玻璃
原相片：谢尔盖·亚历山德罗维奇大公及其妻子黑森和莱茵大公国伊丽莎白公主的合影
高 193.675 毫米　宽 177.8 毫米　厚 22.225 毫米
弗吉尼亚美术馆藏
莉莲·托马斯·普拉特遗赠

　　伊丽莎白公主是亚历山德拉皇后的姐姐，谢尔盖是沙皇亚历山大二世之子、沙皇亚历山大三世的兄弟、沙皇尼古拉二世的叔父。相片下方铭刻的时间（1891～1904 年）代表谢尔盖任莫斯科总督 14 周年。该相框是莉莲·托马斯·普拉特所购的首件法贝热作品。

78

Frame

1899–1908
Fabergé firm
Silver, Mammoth ivory, enamel, glass
Original photograph: Grand Duke Sergii Aleksandrovich and Princess Elisaveta of Hesse and by Rhine
7 5/8" H × 7" W × 7/8" D
Virginia Museum of Fine Arts
Bequest of Lillian Thomas Pratt

　　Princess Elisaveta was Empress Alexandra's older sister, and Sergii was a son of Tsar Alexander II, a brother of Tsar Alexander III, and an uncle of Tsar Nicholas II. The dates (1891–1904) engraved below the photograph refer to the 14th anniversary of Sergii's governorship of the city of Moscow. This frame was the first Fabergé item acquired by Lillian Thomas Pratt.

79

相框

1899～1908 年
法贝热公司
工匠大师：安蒂·内瓦莱南
银镀金、珐琅、冬青、玻璃
照片：英王乔治五世和威尔士王子（即后来的爱德华八世）、沙皇尼古拉二世以及皇储阿列克谢
高 60.325 毫米　宽 234.95 毫米　厚 15.875 毫米
弗吉尼亚美术馆藏
莉莲·托马斯·普拉特遗赠

79

Frame

1899–1908
Fabergé firm
Workmaster: Antti Nevalainen
Silver gilt, enamel, holly, glass
Photo: King George V with the Prince of Wales (future King Edward VIII), Tsar Nicholas II , and Tsesarevich Aleksei
2 3/8" H × 9 1/4" W × 5/8" D
Virginia Museum of Fine Arts
Bequest of Lillian Thomas Pratt

80

相框

1899～1908 年
法贝热公司
工匠大师：安蒂·内瓦莱南
银镀金、珐琅、木、玻璃
照片：女大公奥尔加·亚历山德罗芙娜和希腊王后奥尔加（？）
高 82.55 毫米　宽 101.6 毫米　厚 12.7 毫米
弗吉尼亚美术馆藏
莉莲·托马斯·普拉特遗赠

80

Frame

1899–1908
Fabergé firm
Workmaster: Antti Nevalainen
Silver gilt, enamel, wood, glass
Photo: Grand Duchess Olga Alexandrovna and Queen Olga of Greece (?)
3 1/4" H × 4" W × 1/2" D
Virginia Museum of Fine Arts
Bequest of Lillian Thomas Pratt

81

相框

1899 ~ 1917 年
法贝热公司
黄金、铂、珐琅、赛璐珞（现代）、玻璃、猛犸象牙、水彩
高 60.325 毫米　宽 50.8 毫米　厚 9.525 毫米
弗吉尼亚美术馆藏
莉莲·托马斯·普拉特遗赠

81

Frame

1899–1917
Fabergé firm
Gold, platinum, enamel, celluloid (modern), glass, Mammoth ivory, watercolor
2 3/8" H × 2" W × 3/8" D
Virginia Museum of Fine Arts
Bequest of Lillian Thomas Pratt

82

相框

19 ~ 20 世纪，1947 年以前
未知
法贝热仿制品
银镀金、珐琅、钻石、蓝宝石、玻璃、猛犸象牙
高 152.4 毫米　宽 82.55 毫米
弗吉尼亚美术馆藏
莉莲·托马斯·普拉特遗赠

82

Frame

19th – 20th Century, before 1947
Unknown
Fabergé forgery
Silver gilt, enamel, diamonds, sapphire, glass, Mammoth ivory
6" H × 3 1/4" W
Virginia Museum of Fine Arts
Bequest of Lillian Thomas Pratt

83

相框

19 ~ 20 世纪
俄罗斯
银镀金、黄金、珐琅、赛璐珞（现代）、玻璃
高 95.25 毫米　宽 79.375 毫米　厚 12.7 毫米
弗吉尼亚美术馆藏
莉莲·托马斯·普拉特遗赠

83

Frame

19th – 20th Century
Russian
Silver gilt, gold, enamel, celluloid (modern), glass
3 3/4" H × 3 1/8" W × 1/2" D
Virginia Museum of Fine Arts
Bequest of Lillian Thomas Pratt

84

相框

19 ~ 20 世纪
俄罗斯
银镀金、珐琅、金属
微型画：奥尔加女大公
高 82.55 毫米　宽 66.675 毫米　厚 9.525 毫米
弗吉尼亚美术馆藏
莉莲・托马斯・普拉特遗赠

84

Frame

19th–20th Century
Russian
Miniature: Grand Duchess Olga
Silver gilt, enamel, metal
3 1/4" H × 2 5/8" W × 3/8" D
Virginia Museum of Fine Arts
Bequest of Lillian Thomas Pratt

85

相框

19 ~ 20 世纪
俄罗斯
枫木、银镀金、玻璃、皮革、铁质挂钩
高 311.15 毫米　宽 368.3 毫米　厚 15.875 毫米
弗吉尼亚美术馆藏
莉莲・托马斯・普拉特遗赠

85

Frame

19th–20th Century
Russian
Maple, silver gilt, glass, leather, iron hanging hook
12 1/4" H × 14 1/2" W × 5/8" D
Virginia Museum of Fine Arts
Bequest of Lillian Thomas Pratt

86

相框

约 1900 年
法贝热公司
银镀金、珐琅、水晶、木
高 117.475 毫米　宽 85.725 毫米　厚 12.7 毫米
弗吉尼亚美术馆藏
莉莲·托马斯·普拉特遗赠

86

Frame

ca. 1900
Fabergé firm
Silver gilt, enamel, rock crystal, wood
4 5/8" H × 3 3/8" W × 1/2" D
Virginia Museum of Fine Arts
Bequest of Lillian Thomas Pratt

87

相框

约 1900 年

法贝热公司

鲍文玉（硬绿蛇纹石）、黄金、银镀金、玻璃

高 69.85 毫米　宽 49.213 毫米　厚 9.525 毫米

弗吉尼亚美术馆藏

莉莲·托马斯·普拉特遗赠

87

Frame

ca. 1900

Fabergé firm

Bowenite, gold, silver gilt, glass

2 3/4" H × 1 15/16" W × 3/8" D

Virginia Museum of Fine Arts

Bequest of Lillian Thomas Pratt

88

相框

1908 ~ 1917 年

法贝热公司

银镀金、珐琅、木、玻璃

高 95.25 毫米　宽 98.425 毫米　厚 12.7 毫米

弗吉尼亚美术馆藏

莉莲·托马斯·普拉特遗赠

88

Frame

1908–1917

Fabergé firm

Silver gilt, enamel, wood, glass

3 3/4" H × 3 7/8" W × 1/2"D

Virginia Museum of Fine Arts

Bequest of Lillian Thomas Pratt

89

相框

1908～1917年
法贝热公司
工匠大师：亨瑞克·魏格斯特姆
黄金、银、银镀金、珐琅、玻璃、猛犸象牙
高 66.675 毫米　宽 66.675 毫米　厚 12.7 毫米
弗吉尼亚美术馆藏
莉莲·托马斯·普拉特遗赠

89

Frame

1908–1917
Fabergé firm
Workmaster: Henrik Wigström
Gold, silver, silver gilt, enamel, glass, Mammoth ivory
2 5/8" H × 2 5/8" W × 1/2" D
Virginia Museum of Fine Arts
Bequest of Lillian Thomas Pratt

90

相框

1908～1917年
法贝热公司
青金石、黄金、银、珍珠、玻璃、猛犸象牙
高 88.9 毫米　宽 63.5 毫米　厚 15.875 毫米
弗吉尼亚美术馆藏
莉莲·托马斯·普拉特遗赠

90

Frame

1908–1917
Fabergé firm
Lapis lazuli, gold, silver, pearls, glass, Mammoth ivory
3 1/2" H × 2 1/2" W × 5/8" D
Virginia Museum of Fine Arts
Bequest of Lillian Thomas Pratt

91

相框

1908～1917年
法贝热公司
黄金、珐琅、珍珠、玻璃、猛犸象牙
高 79.375 毫米　宽 66.675 毫米　厚 9.525 毫米
弗吉尼亚美术馆藏
莉莲·托马斯·普拉特遗赠

91

Frame

1908–1917
Fabergé firm
Gold, enamel, pearls, glass, Mammoth ivory
3 1/8" H × 2 5/8" W × 3/8" D
Virginia Museum of Fine Arts
Bequest of Lillian Thomas Pratt

92

相框

1908～1917年
俄罗斯
银镀金、红宝石、蓝宝石、玻璃、天鹅绒
高 139.7 毫米　宽 98.425 毫米　厚 9.525 毫米
弗吉尼亚美术馆藏
莉莲·托马斯·普拉特遗赠

92

Frame

1908–1917
Russian
Silver gilt, rubies, sapphires, glass, velvet
5 1/2" H × 3 7/8" W × 3/8" D
Virginia Museum of Fine Arts
Bequest of Lillian Thomas Pratt

| 93 | 93 |

纪念牌匾

19 世纪
俄罗斯
银、银镀金、珐琅
刻有博尔基救世主基督大教堂的景象（博尔基在乌克兰的哈尔科夫城附近）
高 190.5 毫米　宽 142.875 毫米　厚 25.4 毫米
弗吉尼亚美术馆藏
莉莲·托马斯·普拉特遗赠

Commemorative Plaque

19th Century
Russian
Silver, silver gilt, enamel
Engraved with a view of the Church of Christ the Savior in Borki (close to the city of Kharkov, Ukraine)
7 1/2" H × 5 5/8" W × 1" D
Virginia Museum of Fine Arts
Bequest of Lillian Thomas Pratt

94

圆形袖珍肖像

约 1900 年
俄罗斯
金属、珐琅
直径 57.15 毫米　厚 6.35 毫米
弗吉尼亚美术馆藏
莉莲·托马斯·普拉特遗赠

94

Circular Miniature

ca. 1900
Russian
Metal, enamel
2 1/4" Dia. × 1/4" D
Virginia Museum of Fine Arts
Bequest of Lillian Thomas Pratt

95

沙皇台式肖像

1906 年
弗雷德里希·克希利
黄金、银、珐琅、蓝宝石、钻石、水晶，
微型画：猛犸象牙、水彩
高 117.475 毫米　宽 139.7 毫米　厚 9.525 毫米
弗吉尼亚美术馆藏
莉莲·托马斯·普拉特遗赠

95

Imperial Table Portrait

1906
Friedrich Koechly
Gold, silver, enamel, sapphires, diamonds, rock crystal,
miniature: Mammoth ivory, watercolor
4 5/8" H × 5 1/2" W 3/8" D
Virginia Museum of Fine Arts
Bequest of Lillian Thomas Pratt

水手雕像

约 1900 年
法贝热公司
石英、东陵玉、缟玛瑙、青金石、蓝宝石、黄金
高 120.65 毫米　宽 60.325 毫米　厚 31.75 毫米
弗吉尼亚美术馆藏
莉莲·托马斯·普拉特遗赠

这尊雕像由许多种宝石拼接而成。水手身穿奶白色缟玛瑙制服，面孔则采用棕色东陵玉，眼睛为蓝宝石。他的玛瑙水手帽镶有一圈青金石帽檐，上面刻有镀金铭文。他还打有青金石领结，脚蹬黑曜石鞋子。据悉，法贝热只制作了50尊这样的玉石人物雕像，这就是其中一尊。水手戴的帽子上刻有名字"扎尼察"，这个名字使人联想到沙皇尼古拉二世的叔父米哈伊尔·亚历山大罗维奇大公拥有的一艘皇家游艇（原名"福罗斯"）。1895年，乔治·亚历山大罗维奇大公买下这艘游艇，并为其改名"扎尼察"，意为"夏日闪电"。

法贝热复合宝石人物雕像的灵感，似乎源于为韦廷大公或美第奇大公礼拜堂制作的17世纪萨克森或佛罗伦萨圣像。法贝热作品中的实际人物，却通常基于典型的俄罗斯民族人物或民间传说中的人物形象，但是该公司也会制作特别委托的人物形象和肖像。人物雕像最大的藏家就是沙皇本人，他的小型雕塑都收藏在私人陈列室里。

Statuette of a Sailor

ca. 1900
Fabergé firm
Quartz, aventurine, onyx, lapis lazuli, sapphires, gold
4 3/4" H × 2 3/8" W × 1 1/4" D
Virginia Museum of Fine Arts
Bequest of Lillian Thomas Pratt

This figurine is made of many hardstones. The sailor wears a milky white onyx uniform and has a brownish aventurine face with sapphire eyes. His agate cap has a lapis lazuli brim with a gilt inscription. He wears a lapis tie and obsidian shoes. One of only 50 known hardstone human figures carved by the Fabergé firm, this sailor wears a cap bearing the name *Zarnitsa*. The name links the figure to an imperial yacht owned by Grand Duke Mikhail Aleksandrovich, an uncle of Nicholas II. When Grand Duke Georgii Aleksandrovich acquired the ship in 1895, he renamed it *Zarnitsa*, which means "summer lightning."

The inspiration for Fabergé's composite hardstone figures appears to be from Saxon or Florentine figures of 17th-century saints produced for the chapels of Medici or Wettin princes. The actual figures, however, were often based on typical Russian-national types or characters from folklore, although the firm also produced special commissions and portrait figures. The leading collector of figures was the tsar himself, who kept his statuettes in his private cabinet.

97

水手雕像礼品盒

约 1900 年
枫木、天鹅绒、绸缎、黄铜
高 52.388 毫米　宽 88.9 毫米　厚 152.4 毫米
弗吉尼亚美术馆藏
莉莲·托马斯·普拉特遗赠

97

Presentation Box for Statuette of a Sailor

ca. 1900
Maple, velvet, satin, brass
2 1/16" H × 3 1/2" W × 6" D
Virginia Museum of Fine Arts
Bequest of Lillian Thomas Pratt

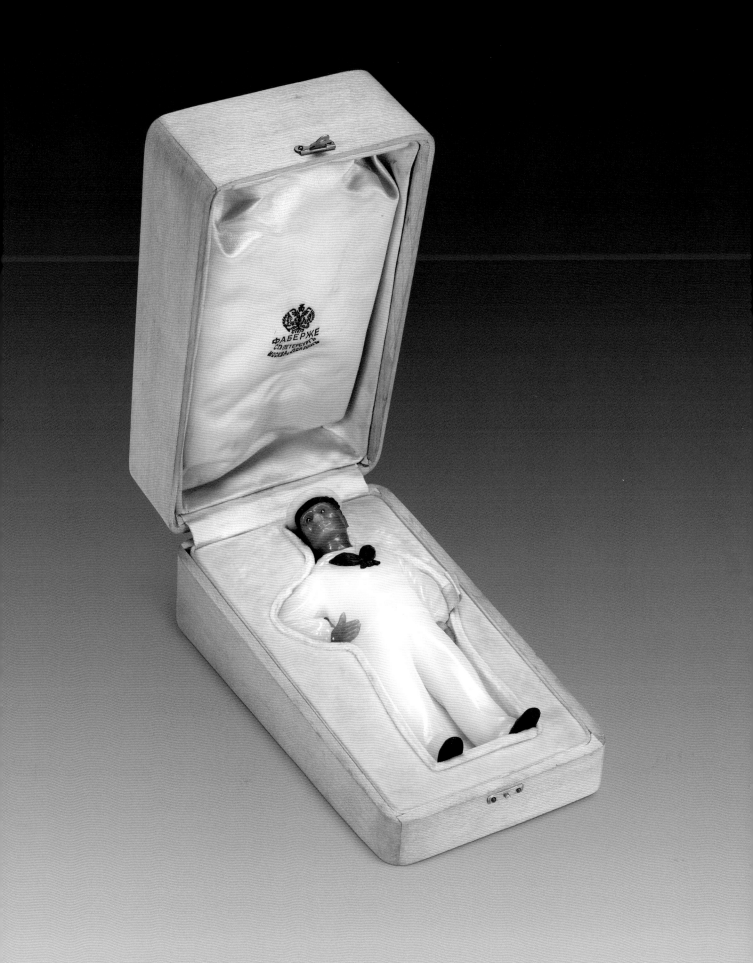

98

鹰

1899 ~ 1908 年
法贝热公司
工匠大师：亨瑞克·魏格斯特姆
玛瑙、黄金、钻石
高 41.275 毫米　长 73.025 毫米　宽 28.575 毫米
弗吉尼亚美术馆藏
莉莲·托马斯·普拉特遗赠

98

Eagle

1899–1908
Fabergé firm
Workmaster: Henrik Wigström
Agate, gold, diamonds
1 5/8" H × 2 7/8" L × 1 1/8" W
Virginia Museum of Fine Arts
Bequest of Lillian Thomas Pratt

99

小鸡

1899 ~ 1908 年
法贝热公司
东陵玉、石英、黄金、红宝石
高 66.675 毫米　宽 50.8 毫米　厚 38.1 毫米
弗吉尼亚美术馆藏
莉莲·托马斯·普拉特遗赠

99

Chick

1899–1908
Fabergé firm
Aventurine, quartz, gold, rubies
2 5/8" H × 2" W × 1 1/2" D
Virginia Museum of Fine Arts
Bequest of Lillian Thomas Pratt

100	101	102
公鸡	公鸡	犀鸟
约 1900 年	约 1900 年	19 ~ 20 世纪
法贝热公司	法贝热公司	被认为是卡地亚作品
玛瑙、黄金、钻石	红玉髓、钻石、黄金	烟晶、黄金、钻石
高 34.925 毫米　长 41.275 毫米	高 34.925 毫米　长 34.925 毫米	高 44.45 毫米　宽 44.45 毫米
宽 15.875 毫米	宽 15.875 毫米	厚 25.4 毫米
弗吉尼亚美术馆藏	弗吉尼亚美术馆藏	弗吉尼亚美术馆藏
莉莲·托马斯·普拉特遗赠	莉莲·托马斯·普拉特遗赠	莉莲·托马斯·普拉特遗赠

100
Rooster

ca. 1900
Fabergé firm
Agate, gold, diamonds
1 3/8" H × 1 5/8" L × 5/8" W
Virginia Museum of Fine Arts
Bequest of Lillian Thomas Pratt

101
Rooster

ca. 1900
Fabergé firm
Carnelian, diamonds, gold
1 3/8" H × 1 3/8" L × 5/8" W
Virginia Museum of Fine Arts
Bequest of Lillian Thomas Pratt

102
Hornbill

19th–20th Century
Attributed to Cartier
Smoky quartz, gold, diamonds
1 3/4" H × 1 3/4" W × 1" D
Virginia Museum of Fine Arts
Bequest of Lillian Thomas Pratt

103

笼中鸣鸟

约 1900 年
法贝热公司
工匠大师：亨瑞克·魏格斯特姆
黄金、银、萤石、软玉、
珍珠、绿松石、钻石
高 101.6 毫米　直径 60.325 毫米
弗吉尼亚美术馆藏
莉莲·托马斯·普拉特遗赠

103

Songbird in a Cage

ca. 1900
Fabergé firm
Workmaster: Henrik Wigström
Gold, silver, fluorite, nephrite,
pearls, turquoise, diamonds
4" H × 2 3/8" Dia.
Virginia Museum of Fine Arts
Bequest of Lillian Thomas Pratt

104

鹭

约 1900 年
法贝热公司
工匠大师：亨瑞克·魏格斯特姆
玛瑙、钻石、黄金
高 114.3 毫米　宽 50.8 毫米　厚 76.2 毫米
弗吉尼亚美术馆藏
莉莲·托马斯·普拉特遗赠

104

Heron

ca. 1900
Fabergé firm
Workmaster: Henrik Wigström
Agate, diamonds, gold
4 1/2" H × 2" W × 3" D
Virginia Museum of Fine Arts
Bequest of Lillian Thomas Pratt

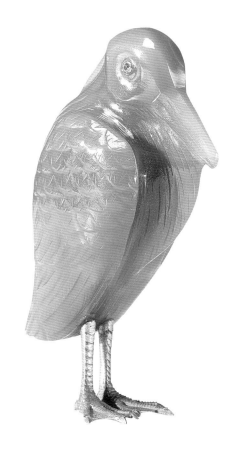

105

鸵鸟

约 1900 年
法贝热公司
玛瑙、翠榴石、黄金、石英岩
高 117.475 毫米　长 69.85 毫米　宽 41.275 毫米
弗吉尼亚美术馆藏
莉莲·托马斯·普拉特遗赠

105

Ostrich

ca. 1900
Fabergé firm
Agate, demantoid garnets, gold, quartzite
4 5/8" H × 2 3/4" L × 1 5/8" W
Virginia Museum of Fine Arts
Bequest of Lillian Thomas Pratt

106

猫头鹰

约 1900 年
法贝热公司
黑曜石、虎眼石、钻石、黄金
高 82.55 毫米　宽 66.675 毫米　厚 79.375 毫米
弗吉尼亚美术馆藏
莉莲·托马斯·普拉特遗赠

106

Owl

ca. 1900
Fabergé firm
Obsidian, tigereyes, diamonds, gold
3 1/4" H × 2 5/8" W × 3 1/8" D
Virginia Museum of Fine Arts
Bequest of Lillian Thomas Pratt

107

鸿雁

约 1900 年
法贝热公司
水晶、黄金、钻石
高 107.95 毫米　长 79.375 毫米　宽 44.45 毫米
弗吉尼亚美术馆藏
莉莲·托马斯·普拉特遗赠

107

Chinese Swan Goose

ca. 1900
Fabergé firm
Rock crystal, gold, diamonds
4 1/4" H × 3 1/8" L × 1 3/4" W
Virginia Museum of Fine Arts
Bequest of Lillian Thomas Pratt

108

鸿雁礼品盒

约 1900 年
冬青、天鹅绒、绸缎、黄铜
高 63.5 毫米　宽 127 毫米　厚 155.575 毫米
弗吉尼亚美术馆藏
莉莲·托马斯·普拉特遗赠

108

Presentation Box for
Chinese Swan Goose

ca. 1900
Holly, velvet, satin, brass
2 1/2" H × 5" W × 6 1/8" D
Virginia Museum of Fine Arts
Bequest of Lillian Thomas Pratt

109	110	111

洪堡企鹅

约 1900 年
法贝热公司
黑曜石、钻石
高 73.025 毫米　宽 44.45 毫米
厚 50.8 毫米
弗吉尼亚美术馆藏
莉莲・托马斯・普拉特遗赠

兔

19 ~ 20 世纪
俄罗斯或法国，可能由卡地亚零售
鲍文玉（硬绿蛇纹石）、红宝石
高 31.75 毫米　长 50.8 毫米
宽 34.925 毫米
弗吉尼亚美术馆藏
莉莲・托马斯・普拉特遗赠

兔形水罐

1899 年以前
法贝热公司
银、黄金、石榴石
高 254 毫米　宽 177.8 毫米
厚 133.35 毫米
弗吉尼亚美术馆藏
莉莲・托马斯・普拉特遗赠

109
Humboldt Penguin

ca. 1900
Fabergé firm
Obsidian, diamonds
2 7/8" H × 1 3/4" W × 2" D
Virginia Museum of Fine Arts
Bequest of Lillian Thomas Pratt

110
Rabbit

19th–20th Century
Russian or French, possibly retailed by Cartier
Bowenite, rubies
1 1/4" H × 2" L × 1 3/8" W
Virginia Museum of Fine Arts
Bequest of Lillian Thomas Pratt

111
Rabbit Pitcher

Before 1899
Fabergé firm
Silver, gold, garnets
10" H × 7" W × 5 1/4" D
Virginia Museum of Fine Arts
Bequest of Lillian Thomas Pratt

112

兔形门铃

1908 ~ 1917 年
法贝热公司
银、石榴石
高 130.175 毫米　宽 66.675 毫米　厚 95.25 毫米
弗吉尼亚美术馆藏
莉莲·托马斯·普拉特遗赠

112

Rabbit Bell Push

1908–1917
Fabergé firm
Silver, garnets
5 1/8" H × 2 5/8" W × 3 3/4" D
Virginia Museum of Fine Arts
Bequest of Lillian Thomas Pratt

113

法国斗牛犬

约 1900 年
被认为是卡地亚作品
烟晶、黄金、银、蓝宝石
高 79.375 毫米　宽 60.325 毫米　厚 47.625 毫米
弗吉尼亚美术馆藏
莉莲·托马斯·普拉特遗赠

114

法国斗牛犬

约 1900 年
被认为是卡地亚作品
烟晶、蓝宝石、黄金
高 34.925 毫米　宽 66.675 毫米　厚 44.45 毫米
弗吉尼亚美术馆藏
莉莲·托马斯·普拉特遗赠

113

French Bulldog

ca. 1900
Attributed to Cartier
Smoky quartz, gold, silver, sapphires
3 1/8" H × 2 3/8" W × 1 7/8" D
Virginia Museum of Fine Arts
Bequest of Lillian Thomas Pratt

114

French Bulldog

ca. 1900
Attributed to Cartier
Smoky quartz, sapphires, gold
1 3/8" H × 2 5/8" W × 1 3/4" D
Virginia Museum of Fine Arts
Bequest of Lillian Thomas Pratt

115	116	117
法国斗牛犬	法国斗牛犬	英国斗牛犬
约 1900 年	约 1900 年	约 1900 年
被认为是卡地亚作品	法贝热公司	法贝热公司
烟晶、红宝石、黄金、珍珠	东陵玉、石英、黄金、珐琅、祖母绿	工匠大师：亨瑞克·魏格斯特姆
高 44.45 毫米　宽 41.275 毫米	高 44.45 毫米　长 57.15 毫米	黑曜石、钻石、黄金、珐琅
厚 28.575 毫米	宽 28.575 毫米	高 50.8 毫米　长 69.85 毫米
弗吉尼亚美术馆藏	弗吉尼亚美术馆藏	宽 31.75 毫米
莉莲·托马斯·普拉特遗赠	小欧内斯特·希尔曼家族捐赠	弗吉尼亚美术馆藏
		莉莲·托马斯·普拉特遗赠

115
French Bulldog

ca. 1900
Attributed to Cartier
Smoky quartz, rubies, gold, pearl
1 3/4" H × 1 5/8" W × 1 1/8" D
Virginia Museum of Fine Arts
Bequest of Lillian Thomas Pratt

116
French Bulldog

ca. 1900
Fabergé firm
Aventurine, quartz, gold, enamel, emeralds
1 3/4" H × 2 1/4" L × 1 1/8" W
Virginia Museum of Fine Arts
Gift of the Estate of Ernest Hillman Jr.

117
English Bulldog

ca. 1900
Fabergé firm
Workmaster: Henrik Wigström
Obsidian, diamonds, gold, enamel
2" H × 2 3/4" L × 1 1/4" W
Virginia Museum of Fine Arts
Bequest of Lillian Thomas Pratt

118

腊肠犬

约 1900 年
法贝热公司
烟玛瑙、钻石
高 50.8 毫米　长 69.85 毫米　宽 38.1 毫米
弗吉尼亚美术馆藏
莉莲·托马斯·普拉特遗赠

　　这件腊肠犬雕像被认为是法贝热真品，是因为它能够完美地装入原配珠宝盒中。没有珠宝盒盛装的动物雕像因为种种原因难以被鉴定为法贝热真品。例如，卡地亚公司就曾出售 1300 件酷似法贝热风格的动物雕像，这些雕像要么购于俄国（一些来自法贝热本人），要么由位于巴黎的法国公司委托制作。而且，一些技艺高超的圣彼得堡工匠与零售商也以专造这类玉石动物雕像广为人知。这其中有许多件即使不属上乘，也质量精良，使得它们与法贝热真品几无差别。而法贝热本人甚至也有可能收购这些雕像进行转卖。

118

Dachshund

ca. 1900
Fabergé firm
Smoky agate, diamonds
2" H × 2 3/4" L × 1 1/2" W
Virginia Museum of Fine Arts
Bequest of Lillian Thomas Pratt

　　This dachshund is deemed a Fabergé object due to the original case into which he fits perfectly. Animals without cases are difficult to verify as true Fabergé creations for a number of reasons. The firm of Cartier, for example, marketed 1,300 Fabergé look-like figures either acquired in Russia (some of them from Fabergé himself) or commissioned by the French firm in Paris. Furthermore, a number of skilled Russian craftsmen or retailers are now known to have specialized in such hardstone animals in St. Petersburg. Many of these are of good, if not excellent, quality, rendering them virtually indistinguishable from those of Fabergé. Indeed, Fabergé may have even acquired some of these figures to sell himself.

119

腊肠犬礼品盒

约 1900 年
枫木、天鹅绒、绸缎、黄铜
高 44.45 毫米　宽 111.125 毫米
厚 82.55 毫米
弗吉尼亚美术馆藏
莉莲·托马斯·普拉特遗赠

119

Presentation Box for Dachshund

ca. 1900
Maple, velvet, satin, brass
1 3/4" H × 4 3/8" W × 3 1/4" D
Virginia Museum of Fine Arts
Bequest of Lillian Thomas Pratt

120

腊肠犬

约 1900 年
法贝热公司
玛瑙、红宝石
高 31.75 毫米　长 76.2 毫米
宽 22.225 毫米
弗吉尼亚美术馆藏
莉莲·托马斯·普拉特遗赠

121

猪

19～20 世纪
俄罗斯或法国，可能由卡地亚零售
青金石、钻石
高 44.45 毫米　长 76.2 毫米
宽 31.75 毫米
弗吉尼亚美术馆藏
莉莲·托马斯·普拉特遗赠

120

Dachshund

ca. 1900
Fabergé firm
Agate, rubies
1 1/4" H × 3" L × 7/8" W
Virginia Museum of Fine Arts
Bequest of Lillian Thomas Pratt

121

Pig

19th–20th Century
Russian or French, possibly retailed by Cartier
Lapis lazuli, diamonds
1 3/4" H × 3" L × 1 1/4" W
Virginia Museum of Fine Arts
Bequest of Lillian Thomas Pratt

122

象

19～20世纪，1947年以前
未知
法贝热仿制品
银
高 38.1 毫米　宽 25.4 毫米
厚 50.8 毫米
弗吉尼亚美术馆藏
莉莲·托马斯·普拉特遗赠

123

象

19～20世纪，1947年以前
未知
法贝热仿制品
铜鎏金
高 38.1 毫米　宽 25.4 毫米
厚 50.8 毫米
弗吉尼亚美术馆藏
莉莲·托马斯·普拉特遗赠

124

象

19～20世纪
俄罗斯或法国，可能由卡地亚零售
软玉、银、钻石
高 76.2 毫米　宽 50.8 毫米
厚 88.9 毫米
弗吉尼亚美术馆藏
莉莲·托马斯·普拉特遗赠

122

Elephant

19th–20th Century, before 1947
Unknown
Fabergé forgery
Silver
1 1/2" H × 1" W × 2" D
Virginia Museum of Fine Arts
Bequest of Lillian Thomas Pratt

123

Elephant

19th–20th Century, before 1947
Unknown
Fabergé forgery
Gilt brass
1 1/2" H × 1" W × 2" D
Virginia Museum of Fine Arts
Bequest of Lillian Thomas Pratt

124

Elephant

19th–20th Century
Russian or French, possibly retailed by Cartier
Nephrite, silver, diamonds
3" H × 2" W × 3 1/2" D
Virginia Museum of Fine Arts
Bequest of Lillian Thomas Pratt

125

英国山楂

1899～1917 年
法贝热公司
软玉、条带状方解石、黄金、
玉髓、东陵玉、珊瑚
高 131.763 毫米　宽 95.25 毫米　厚 66.675 毫米
弗吉尼亚美术馆藏
莉莲·托马斯·普拉特遗赠

125

English Hawthorn

1899–1917
Fabergé firm
Nephrite, banded calcite, gold,
chalcedony, aventurine, coral
5 3/16" H × 3 3/4" W × 2 5/8" D
Virginia Museum of Fine Arts
Bequest of Lillian Thomas Pratt

126

金莲花

约 1900 年
法贝热公司
黄金、珐琅、软玉、水晶
高 127 毫米　宽 120.65 毫米　厚 79.375 毫米
弗吉尼亚美术馆藏
莉莲·托马斯·普拉特遗赠

126

Globeflowers

ca. 1900
Fabergé firm
Gold, enamel, nephrite, rock crystal
5" H × 4 3/4" W × 3 1/8" D
Virginia Museum of Fine Arts
Bequest of Lillian Thomas Pratt

127

罂粟花

19～20世纪，1947年以前
未知
法贝热仿制品
银镀金、玉髓、软玉、玛瑙、
黄金、蓝宝石、黄玉、钻石
高 247.65 毫米　宽 101.6 毫米　厚 88.9 毫米
弗吉尼亚美术馆藏
莉莲·托马斯·普拉特遗赠

127

Poppy

19th–20th Century, before 1947
Unknown
Fabergé forgery
Silver gilt, chalcedony, nephrite,
agate, gold, sapphires, topaz, diamond
9 3/4" H × 4" W × 3 1/2" D
Virginia Museum of Fine Arts
Bequest of Lillian Thomas Pratt

128

报春花

19～20世纪，1947年以前
未知
法贝热仿制品
红玉髓、软玉、水晶、黄金、钻石
高101.6毫米　宽76.2毫米
厚76.2毫米
弗吉尼亚美术馆藏
莉莲·托马斯·普拉特遗赠

128

Primrose

19th–20th Century, before 1947
Unknown
Fabergé forgery
Carnelian, nephrite, rock crystal,
gold, diamonds
4" H × 3" W × 3" D
Virginia Museum of Fine Arts
Bequest of Lillian Thomas Pratt

129

铃兰

19 ~ 20 世纪，1947 年以前
未知
法贝热仿制品
磨砂水晶、软玉、黄金
高 152.4 毫米　宽 69.85 毫米
厚 44.45 毫米
弗吉尼亚美术馆藏
莉莲·托马斯·普拉特遗赠

129

Lilies of the Valley

19th–20th Century, before 1947
Unknown
Fabergé forgery
Frosted rock crystal, nephrite, gold
6" H × 2 3/4" W × 1 3/4" D
Virginia Museum of Fine Arts
Bequest of Lillian Thomas Pratt

130

金凤花

19 ~ 20 世纪，1947 年以前
未知
法贝热仿制品
玛瑙、软玉、青金石、鲍文玉（硬绿蛇纹石）、
黄金、钻石、银
高 146.05 毫米　宽 76.2 毫米
厚 57.15 毫米
弗吉尼亚美术馆藏
莉莲·托马斯·普拉特遗赠，

130

Buttercup

19th–20th Century, before 1947
Unknown
Fabergé forgery
Agate, nephrite, lapis lazuli,
bowenite, gold, diamonds, silver
5 3/4" H × 3" W × 2 1/4" D
Virginia Museum of Fine Arts
Bequest of Lillian Thomas Pratt

131

蓝铃花

19～20 世纪，1947 年以前
未知
法贝热仿制品
玉髓、软玉、玛瑙、黄金、翠榴石
高 161.925 毫米　宽 57.15 毫米　厚 44.45 毫米
弗吉尼亚美术馆藏
莉莲·托马斯·普拉特遗赠

　　在弗吉尼亚美术馆普拉特藏品中的 22 件花卉作品里，唯有一件真品矢车菊花枝购于 1934 年，其余均由普拉特夫人于 1939 年 1 月 9 日至 1940 年 1 月 9 日期间从阿曼德·哈默手上购入。大约从 1930 年至 1939 年，这位美国商业大亨阿曼德·哈默及其兄弟维克多成为苏联官员在美国出售俄罗斯皇室珍宝的独家代理。哈默出售给普拉特夫人的花卉作品中有 15 件极有可能是赝品。这些赝品风格各异：有些工艺精制，可能出自法国；而另一些较为粗糙，可能是德国制造。但是，这些赝品都打上了法贝热和魏格斯特姆的印记，当然这些印记都是后来被加上去的。

131

Bluebells

19th – 20th Century, before 1947
Unknown
Fabergé forgery
Chalcedony, nephrite, agate, gold, demantoid garnets
6 3/8" H × 2 1/4" W × 1 3/4" D
Virginia Museum of Fine Arts
Bequest of Lillian Thomas Pratt

　　Of the 22 flowers in VMFA's Pratt collection, all but one — the genuine *Cornflower Spray* purchased in 1934 — were acquired by Mrs. Pratt between January 9, 1939, and January 9, 1940, from Armand Hammer. From roughly 1930 to 1939, this American business tycoon and his brother Victor represented Soviet officials as sole agents for the disposal of Russian imperial treasure in the United States. Fifteen of the flowers Hammer sold to Mrs. Pratt are most probably forgeries. Their styles vary: some are more finely crafted, perhaps denoting a French origin; others are cruder, possibly indicating a German provenance. However, they all share the Fabergé and Wigström hallmarks that were added later.

132

三色堇

19 ～ 20 世纪，1947 年以前

未知

法贝热仿制品

紫水晶、软玉、水晶、黄金、钻石

高 133.35 毫米　宽 76.2 毫米　厚 95.25 毫米

弗吉尼亚美术馆藏

莉莲・托马斯・普拉特遗赠

132

Pansies

19th–20th Century, before 1947

Unknown

Fabergé forgery

Amethysts, nephrite, rock crystal, gold, diamonds

5 1/4" H × 3" W × 3 3/4" D

Virginia Museum of Fine Arts

Bequest of Lillian Thomas Pratt

133

紫苑

19 ～ 20 世纪，1947 年以前

未知

法贝热仿制品

玉髓、软玉、翠榴石、黄金

高 149.225 毫米　宽 69.85 毫米　厚 57.15 毫米

弗吉尼亚美术馆藏

莉莲・托马斯・普拉特遗赠

133

Aster

19th–20th Century, before 1947

Unknown

Fabergé forgery

Chalcedony, nephrite, demantoid garnets, gold

5 7/8" H × 2 3/4" W × 2 1/4" D

Virginia Museum of Fine Arts

Bequest of Lillian Thomas Pratt

134

郁金香

19～20世纪，1947年以前
未知
法贝热仿制品
紫水晶、软玉、玛瑙、
黄金、银镀金、钻石
高 209.55 毫米　宽 104.775 毫米
厚 60.325 毫米
弗吉尼亚美术馆藏
莉莲·托马斯·普拉特遗赠

134

Tulip

19th–20th Century, before 1947
Unknown
Fabergé forgery
Amethyst, nephrite, agate,
gold, silver gilt, diamonds
8 1/4" H × 4 1/8" W × 2 3/8" D
Virginia Museum of Fine Arts
Bequest of Lillian Thomas Pratt

135

马鞭草

19~20世纪，1947年以前
未知
法贝热仿制品
玉髓、软玉、玛瑙、黄金、钻石、银
高 133.35 毫米　宽 95.25 毫米
厚 85.725 毫米
弗吉尼亚美术馆藏
莉莲·托马斯·普拉特遗赠

135

Verbena

19th–20th Century, before 1947
Unknown
Fabergé forgery
Chalcedony, nephrite, agate,
gold, diamonds, silver
5 1/4" H × 3 3/4" W × 3 3/8" D
Virginia Museum of Fine Arts
Bequest of Lillian Thomas Pratt

136

三色堇

1908 ~ 1917 年
法贝热公司
软玉、水晶、蓝宝石、钻石、珐琅
高 95.25 毫米　宽 57.15 毫米
厚 34.925 毫米
弗吉尼亚美术馆藏
莉莲·托马斯·普拉特遗赠

136

Pansy

1908–1917
Fabergé firm
Nephrite, rock crystal,
sapphires, diamonds, enamel
3 3/4" H × 2 1/4" W × 1 3/8" D
Virginia Museum of Fine Arts
Bequest of Lillian Thomas Pratt

137

紫罗兰

约 1908 年
法贝热公司
黄金、珐琅、钻石、软玉、水晶
高 85.725 毫米　宽 34.925 毫米　厚 28.575 毫米
弗吉尼亚美术馆藏
莉莲·托马斯·普拉特遗赠

138

紫罗兰礼品盒

约 1908 年
枫木、天鹅绒、绸缎、黄铜
高 60.325 毫米　长 158.75 毫米
宽 82.55 毫米
弗吉尼亚美术馆藏
莉莲·托马斯·普拉特遗赠

137

Violet

ca. 1908
Fabergé firm
Gold, enamel, diamond, nephrite, rock crystal
3 3/8" H × 1 3/8" W × 1 1/8" D
Virginia Museum of Fine Arts
Bequest of Lillian Thomas Pratt

138

Presentation Box for Violet

ca. 1908
Maple, velvet, satin, brass
2 3/8" H × 6 1/4" L × 3 1/4" W
Virginia Museum of Fine Arts
Bequest of Lillian Thomas Pratt

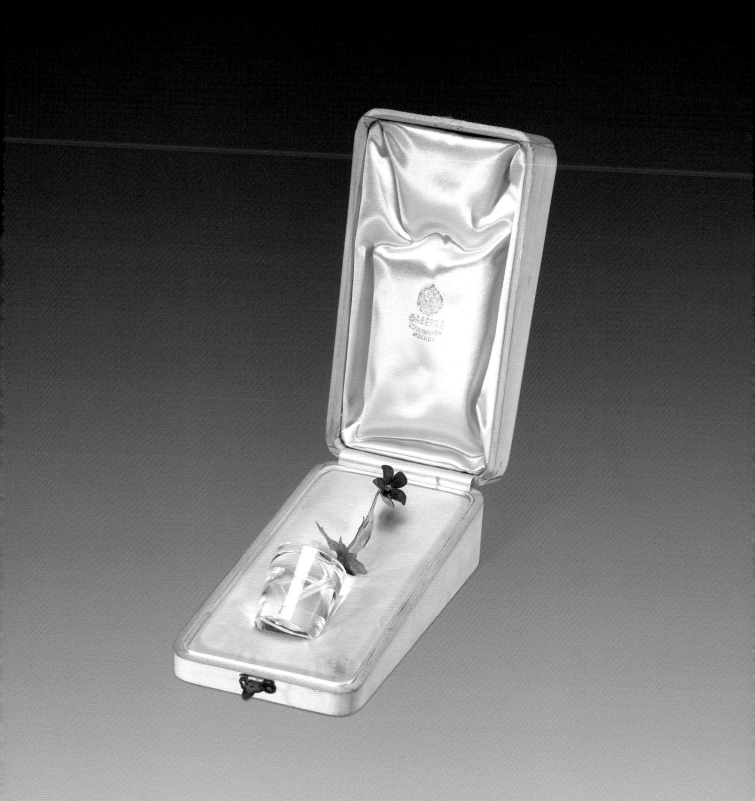

日常的奢华
Everyday Opulence

法贝热根据不同的创作主体，大胆使用陶瓷、玻璃、钢铁、木材、小粒珍珠等材料。除了材料和工艺，法贝热特别注重设计，他的作品体现出哥特式、文艺复兴、巴洛克、新艺术等多种风格，有的作品甚至有强烈的现代感，预见了日后的简约风格。除了熟练而精准地运用传统色调，法贝热还启用了黄、紫红、橙红和各种各样的绿色——总共有超过140种全新的颜色。即使是普通的日常用品，也被打造得精致奢华，充满了浓郁的艺术气息。

Fabergé made bold use of porcelain, glass, iron and steel, wood, tiny pearls and other materials, depending on subjects of creation. In addition to material and craftsmanship, he gave particular importance to design. His works display a great variety of styles including Gothic, Renaissance, Baroque, and Art Nouveau. Some works have so strong a modern feel as to have foreshadowed the simplistic styles that emerged in later days. In addition to traditional colors which he used masterfully, Fabergé also began using yellow, purple red, orange red, and all shades of green — more than 140 new hues altogether. Even ordinary everyday items were also made exquisite and full of a strong air of art.

139	139
加冕纪念杯	Coronation Beaker

1896 年
俄罗斯
珐琅、铁、镀金
高 104.775 毫米　直径 95.25 毫米
弗吉尼亚美术馆藏
布莱克默夫人捐赠，以纪念其夫
莫里斯·内维尔·布莱克默少校

1896
Russian
Enamel, iron, gilding
4 1/8" H × 3 3/4" Dia.
Virginia Museum of Fine Arts
Gift of Mrs. M. N. Blakemore in memory of her husband,
Major Maurice Neville Blakemore

140

袖珍浅杯

1899 年以前
法贝热公司
工匠大师：埃里克·科林
软玉、黄金、蓝宝石、珍珠
高 38.1 毫米　宽 69.85 毫米　厚 41.275 毫米
弗吉尼亚美术馆藏
小欧内斯特·希尔曼家族捐赠

140

Miniature Tazza

Before 1899
Fabergé firm
Workmaster: Erik Kollin
Nephrite, gold, sapphires, pearls
1 1/2" H × 2 3/4" W × 1 5/8" D
Virginia Museum of Fine Arts
Gift of the Estate of Ernest Hillman Jr.

141

三叶形杯子

1899年以前
法贝热公司
工匠大师：埃里克·科林
黄金
高 28.575 毫米　宽 133.35 毫米
厚 133.35 毫米
弗吉尼亚美术馆藏
莉莲·托马斯·普拉特遗赠

141

Trefoil Cup

Before 1899
Fabergé firm
Workmaster: Erik Kollin
Gold
1 1/8" H × 5 1/4" W × 5 1/4" D
Virginia Museum of Fine Arts
Bequest of Lillian Thomas Pratt

142

纺车形杯

1899 年以前
法贝热公司
工匠大师：埃里克·科林
水晶、黄金、红宝石、钻石
高 50.8 毫米　长 85.725 毫米
宽 63.5 毫米
弗吉尼亚美术馆藏
莉莲·托马斯·普拉特遗赠

142

Charka

Before 1899
Fabergé firm
Workmaster: Erik Kollin
Rock crystal, gold, rubies, diamonds
2" H × 3 3/8" L × 2 1/2" W
Virginia Museum of Fine Arts
Bequest of Lillian Thomas Pratt

143

纺车形杯

1899～1908 年
法贝热公司
工匠大师：亨瑞克·魏格斯特姆
血星石、黄金、珐琅、红宝石
高 34.925 毫米　宽 66.675 毫米
厚 53.975 毫米
弗吉尼亚美术馆藏
乔治·露西藏品；爱丽丝和路易斯·尼尔森捐赠，
以庆祝弗吉尼亚美术馆建馆 75 周年

143

Charka

1899–1908
Fabergé firm
Workmaster: Henrik Wigström
Bloodstone, gold, enamel, ruby
1 3/8" H × 2 5/8" W × 2 1/8" D
Virginia Museum of Fine Arts
Collection of Georges Lurcy;
Gift of Alice and Lewis Nelson in celebration
of VMFA's 75th Anniversary

144

纺车形杯礼品盒

1899～1908 年
枫木、天鹅绒、绸缎、黄铜
高 41.275 毫米　宽 88.9 毫米　厚 76.2 毫米
弗吉尼亚美术馆藏
乔治·露西藏品；爱丽丝和路易斯·尼尔森捐赠，
以庆祝弗吉尼亚美术馆建馆 75 周年

144

Presentation Box for Charka

1899–1908
Maple, velvet, satin, brass
1 5/8" H × 3 1/2" W × 3" D
Virginia Museum of Fine Arts
Collection of Georges Lurcy;
Gift of Alice and Lewis Nelson in celebration
of VMFA's 75th Anniversary

纪念版长柄船形酒杯

1899 ~ 1908 年
法贝热公司
银、黄金、绿玉髓、紫水晶
高 381 毫米　长 723.9 毫米　宽 311.15 毫米
弗吉尼亚美术馆藏
杰罗姆和丽塔·甘斯俄罗斯珐琅藏品

长柄船形酒杯是俄罗斯传统饮用器皿，通常为单柄船形，器型不大。早期的长柄船形酒杯为木质。16 世纪以来，人们开始用黄金和白银制作长柄船形酒杯。这只纪念版长柄船形酒杯是现存的法贝热家族制作的银器杰作之一。

这只巨型俄罗斯老式长柄酒杯的船头位置为一群冲锋陷阵的博加提里。博加提里即中世纪时期效忠于基辅弗拉基米尔大公的勇士，其中最著名的三位勇士为伊利亚·穆罗梅茨、阿廖沙·波波维奇和多布雷尼亚·尼基季奇。这些勇士和其他一些传奇人物是老式俄罗斯复兴时期珐琅画上的常见人物形象。这只长柄船形酒杯的杯身是传统的俄罗斯主题与新艺术主义设计的融合，并镶有大颗西伯利亚半宝石。

法贝热工匠中还有擅长制作各式银器且手艺精湛的银匠。俄国在革命前夕，金银器均以兹罗提尼克为纯度单位。兹罗提尼克一词源于俄语，意为黄金。纯银的兹罗提尼克含量为 96%。法贝热银器中最常用的纯度比例为 84%、88% 和 91%。

179

145

Monumental Kovsh

1899–1908
Fabergé firm
Silver, gold, chrysoprase, amethysts
15" H × 28 1/2" L × 12 1/4" W
Virginia Museum of Fine Arts
Jerome and Rita Gans Collection of Russian Enamel

A *kovsh* is a traditional Russian drinking vessel, usually a small, boat-shape with a single handle. Early versions were made of wood, but by the 16th century they began to be made of gold and silver. This monumental *kovsh* is one of the finest pieces of silver made by the House of Fabergé in existence.

The prow of this huge sloping Old-Russian *kovsh* is modeled with a charging group of *bogatyri*, the medieval heroic warriors who served Grand Prince Vladimir in Kiev. The most famous warriors were Il'ia Muromets, Alësha Popovich, and Dobrynia Nikitich. These and other legendary figures were often subjects of paintings on enamel during the Old-Russian Revival. The body of this *kovsh* combines traditional Russian motifs with Art Nouveau designs, and is set with large semiprecious Siberian stones.

Fabergé's workmasters included many expert silversmiths who made a wide variety of objects. In prerevolutionary Russia, both gold and silver standards were expressed in zolotniki, although the word actually comes from the Russian zoloto, meaning gold. Fine or pure silver contained 96 zolotniki. The most frequently used proportions for Fabergé silver were 84, 88, and 91.

146

长柄船形酒杯

1899 年以前
法贝热公司
工匠大师：埃里克·科林
黄金
高 41.275 毫米　长 88.9 毫米
宽 60.325 毫米
弗吉尼亚美术馆藏
莉莲·托马斯·普拉特遗赠

146

Kovsh

Before 1899
Fabergé firm
Workmaster: Erik Kollin
Gold
1 5/8" H × 3 1/2" L × 2 3/8" W
Virginia Museum of Fine Arts
Bequest of Lillian Thomas Pratt

147

长柄船形酒杯

1899 ~ 1908 年
法贝热公司
水晶、黄金、银、红宝石、钻石、珍珠
高 63.5 毫米　长 142.875 毫米　宽 76.2 毫米
弗吉尼亚美术馆藏
莉莲·托马斯·普拉特遗赠

147

Kovsh

1899–1908
Fabergé firm
Rock crystal, gold, silver, rubies, diamonds, pearl
2 1/2" H × 5 5/8" L × 3" W
Virginia Museum of Fine Arts
Bequest of Lillian Thomas Pratt

148

长柄船形酒杯

1908 ~ 1917 年
玛丽亚·塞曼诺娃
银镀金、珐琅、西伯利亚硬岩
高 171.45 毫米　长 400.05 毫米　厚 254 毫米
弗吉尼亚美术馆藏
杰罗姆和丽塔·甘斯俄罗斯珐琅藏品

148

Kovsh

1908–1917
Maria Semenova
Silver gilt, enamel, Siberian hardstones
6 3/4" H × 15 3/4" L × 10" D
Virginia Museum of Fine Arts
Jerome and Rita Gans Collection of Russian Enamel

149

袖珍长柄船形酒杯

1899 ~ 1908 年
法贝热公司
工匠大师：安蒂·内瓦莱南
银镀金、珐琅、月长石
高 44.45 毫米　长 104.775 毫米　宽 73.025 毫米
弗吉尼亚美术馆藏
莉莲·托马斯·普拉特遗赠

149

Miniature Kovsh

1899–1908
Fabergé firm
Workmaster: Antti Nevalainen
Silver gilt, enamel, moonstone
1 3/4" H × 4 1/8" L × 2 7/8" W
Virginia Museum of Fine Arts
Bequest of Lillian Thomas Pratt

150

赞颂杯

1899 ~ 1908 年
费多尔·吕克特
银镀金、珐琅、石榴石
高 234.95 毫米　直径 219.075 毫米
弗吉尼亚美术馆藏
杰罗姆和丽塔·甘斯俄罗斯珐琅藏品

　　这只掐丝珐琅三柄杯采用老式俄罗斯风格。杯身的三幅画分别描绘了一个蓄着胡子、身穿蓝色短袍、手举杯子的基辅罗斯贵族；一位手持托盘的女贵族，盘中摆有一个瓶子和一个杯子；还有一位弹奏乐器的音乐家。

150

Loving Cup

1899–1908
Fedor Rückert
Silver gilt, enamel, garnets
9 1/4" H × 8 5/8" Dia.
Virginia Museum of Fine Arts
Jerome and Rita Gans Collection of Russian Enamel

　　This cup is decorated with cloisonné enamels in the Old-Russian style. The bowl has three panels that depict a boyar (the highest rank of nobility in Kievan Russia) with a beard and a blue tunic holding a cup aloft; a boyarina (the female form of boyar) offering a tray with a carafe and a cup; and a musician with an instrument.

151

果酒杯

约 1900 年
法贝热公司
工匠大师：朱利叶斯·拉波波特
银镀金、珐琅、蓝宝石、祖母绿、红宝石、
石榴石、蓝黄玉、珍珠
高 142.875 毫米　直径 155.575 毫米
弗吉尼亚美术馆藏
莉莲·托马斯·普拉特遗赠

古代俄罗斯制作了许多具有鲜明民族风格的器皿，其中就包括果酒杯"bratina"，意为"兄弟杯"。这种浑圆的果酒杯通常采用金或银的材质制作，杯中盛满蜂蜜酒，并从长兄传至幼弟。这是一件稀有的莫斯科风格作品，以掐丝珐琅和錾胎珐琅装饰，于圣彼得堡制作。

151

Bratina

ca. 1900
Fabergé firm
Workmaster: Julius Rappoport
Silver gilt, enamel, sapphires, emeralds, rubies, garnets, blue topaz, pearls
5 5/8" H × 6 1/8" Dia.
Virginia Museum of Fine Arts
Bequest of Lillian Thomas Pratt

Ancient Russia produced many vessels with a distinct national style, including the *bratina*, which means "brother cup." These rounded cups, usually made of gold or silver, were filled with mead and passed from the oldest brother to the youngest. This is a rare Moscow-type object in cloisonné and champlevé enamel decoration that was made in St. Petersburg.

152

广口杯

1908～1917 年
法贝热公司
银、黄金、珐琅、钻石
高 66.675 毫米　直径 53.975 毫米
弗吉尼亚美术馆藏
莉莲·托马斯·普拉特遗赠

152

Beaker

1908–1917
Fabergé firm
Silver, gold, enamel, diamonds
2 5/8" H × 2 1/8" Dia.
Virginia Museum of Fine Arts
Bequest of Lillian Thomas Pratt

153

杯子

1899～1908年
费多尔·吕克特
银镀金、珐琅
高 254 毫米　直径 95.25 毫米
弗吉尼亚美术馆藏
杰罗姆和丽塔·甘斯俄罗斯珐琅藏品

153

Cup

1899–1908
Fedor Rückert
Silver gilt, enamel
10" H × 3 3/4" Dia.
Virginia Museum of Fine Arts
Jerome and Rita Gans Collection
of Russian Enamel

154

杯子

1908 ~ 1917 年
法贝热公司
工匠大师：亨瑞克·魏格斯特姆
软玉、银镀金、红宝石、蓝宝石
高 88.9 毫米　直径 101.6 毫米
弗吉尼亚美术馆藏
莉莲·托马斯·普拉特遗赠

154

Cup

1908–1917
Fabergé firm
Workmaster: Henrik Wigström
Nephrite, silver gilt, rubies, sapphires
3 1/2" H × 4" Dia.
Virginia Museum of Fine Arts
Bequest of Lillian Thomas Pratt

155

带盖杯

1887～1917 年
工匠大师：费多尔·吕克特
银镀金、珐琅
杯子：高 279.4 毫米　直径 127 毫米
盖：高 101.6 毫米　直径 82.55 毫米
弗吉尼亚美术馆藏
杰罗姆和丽塔·甘斯俄罗斯珐琅藏品

155

Cup and Cover

1887–1917
Workmaster: Fedor Rückert
Silver gilt, enamel
Cup: 11" H × 5" Dia.
Cover: 4" H × 3 1/4" Dia.
Virginia Museum of Fine Arts
Jerome and Rita Gans Collection
of Russian Enamel

156

带盖啤酒杯

1890 年
帕维尔·奥夫钦尼科夫
银镀金、珐琅
高 206.375 毫米　宽 133.35 毫米　厚 101.6 毫米
弗吉尼亚美术馆藏
杰罗姆和丽塔·甘斯俄罗斯珐琅藏品

该酒杯的设计原型来自早前收藏于克里姆林宫军械库博物馆的一只土耳其酒杯，这件土耳其作品由于费多尔·索恩采夫在其著名的六卷本《俄罗斯国家文物，1849～1853 年》中的图示，才被俄罗斯的银器匠们广为熟知。

156

Tankard and Cover

1890
Pavel Ovchinnikov
Silver gilt, enamel
8 1/8" H × 5 1/4" W × 4" D
Virginia Museum of Fine Arts
Jerome and Rita Gans Collection of Russian Enamel

The design of this tankard is based on a Turkish prototype formerly in the Kremlin Armory Museum, known to Russian silversmiths by the drawings published in Fedor Solntsev's six-volume work titled *Antiquities of the Russian State 1849–1853*.

157

带盖杯

1899年以前
法贝热公司
工匠大师：米哈伊尔·佩尔欣
软玉、银镀金、黄金、红宝石、蓝宝石
高 285.75 毫米　直径 139.7 毫米
弗吉尼亚美术馆藏
莉莲·托马斯·普拉特遗赠

157

Cup and Cover

Before 1899
Fabergé firm
Workmaster: Mikhail Perkhin
Nephrite, silver gilt, gold, rubies, sapphires
11 1/4" H × 5 1/2" Dia.
Virginia Museum of Fine Arts
Bequest of Lillian Thomas Pratt

158

带盖杯

1899 ~ 1908 年
法贝热公司
工匠大师：米哈伊尔·佩尔欣
软玉、银镀金、红宝石
高 136.525 毫米　直径 63.5 毫米
弗吉尼亚美术馆藏
莉莲·托马斯·普拉特遗赠

158

Cup and Cover

1899–1908
Fabergé firm
Workmaster: Mikhail Perkhin
Nephrite, silver gilt, ruby
5 3/8" H × 2 1/2" Dia.
Virginia Museum of Fine Arts
Bequest of Lillian Thomas Pratt

面包和盐之盛器

1888 年
I.P. 赫列勃尼科夫公司
银、银镀金、珐琅
高 600.075 毫米　宽 533.4 毫米　厚 50.8 毫米
弗吉尼亚美术馆藏
莉莲·托马斯·普拉特遗赠

摆出面包和盐作为迎接客人的礼仪，是俄罗斯悠久而光荣的传统。这件被称为"面包和盐之盛器"的摆盘中央饰有珐琅制作的赫尔松省盾徽，并刻有铭文："忠诚的赫尔松市省地方自治组织敬赠。"地方自治组织即选举产生的主管教育、交通和卫生的地方自治机构。盘子的镶边上饰有珐琅制作的、加冕的沙皇亚历山大三世姓名首字母标记，还雕刻有三幅点缀涡卷叶纹的地形风景图，分别配有题刻："教育：乡村农业学校；运输：横跨因库列茨河的乡村浮桥；以及卫生：安纳耶夫市医院。"

这只"面包和盐之盛器"是 1888 年沙皇亚历山大三世和妻子玛利亚·费奥多罗芙娜皇后访问赫尔松市省时收到的礼物。沙皇夫妇曾无数次到俄罗斯各地访问，并收到几百只这样的"面包和盐"的摆盘，这只是其中之一。这些盘子最初被展示在圣彼得堡冬宫的前厅和尼古拉斯大厅。

I. P. 赫列勃尼科夫公司是 19 世纪中期至 20 世纪初俄罗斯著名的银器和珠宝制造商之一。1869 年或 1870 年，伊凡·彼得罗维奇·赫列勃尼科夫在莫斯科创立了这家公司，并于 1875 年获得皇家认证。1881 年赫列勃尼科夫去世后，公司由他的儿子们接管。1888 年，赫列勃尼科夫的工坊和店铺合并为 I. P. 赫列勃尼科夫公司。1877 年，I. P. 赫列勃尼科夫在圣彼得堡拥有一家店铺。1881 年至 1914 年期间，该公司的雇员人数在 150 人至 250 人左右。

Bread-and-Salt Dish

1888
I.P.Khlebnikov and Company
Silver, silver gilt, enamel
23 5/8" H × 21" W × 2" D
Virginia Museum of Fine Arts
Bequest of Lillian Thomas Pratt

Welcoming guests with a ceremonial presentation of bread and salt was an ancient and honored tradition in Russia. The center of this presentation dish, known as a bread-and-salt plate, is decorated with the enameled coat of arms of the Kherson province. The inscription reads: "From the loyal *zemstvo* of the Kherson *Guberniia*." A zemstvo was an elected body of local self-government in charge of education, transport, and health. The border of the dish is featured with the crowned and enameled cipher of Tsar Alexander III and has three engraved topographical views interspersed with panels of scrolling foliage. The views are inscribed: *Education: Village Agricultural School; Transport: Village Ponton Bridge across River Inguletz; and Health: Hospital in City of Anayev.*

This bread-and-salt plate was presented to Tsar Alexander III and his wife Maria Feodorovna in 1888 on the occasion of their visit to the Kherson province. It is one of hundreds of such bread-and-salt plates presented to the tsar and his wife at their numerous visits throughout Russia. These plates were originally displayed in the Ante-chamber and the Nicholas Hall of the Winter Palace in St. Petersburg.

The firm of I. P. Khlebnikov and Company was one of the leading silver and jewelry manufacturers in Russia from the mid 19th to the early 20th century. Ivan Petrovich Khlebnikov opened his Moscow firm in 1869 or 1870, and received the imperial warrant in 1875. After his death in 1881, the firm was inherited by his sons. In 1888 Khlebnikov's workshops and stores were combined to form I. P. Khlebnikov and Company. By 1877, I. P. Khlebnikov also had a shop in St. Petersburg. Between 1881 and 1914 the company employed around 150 to 250 workers.

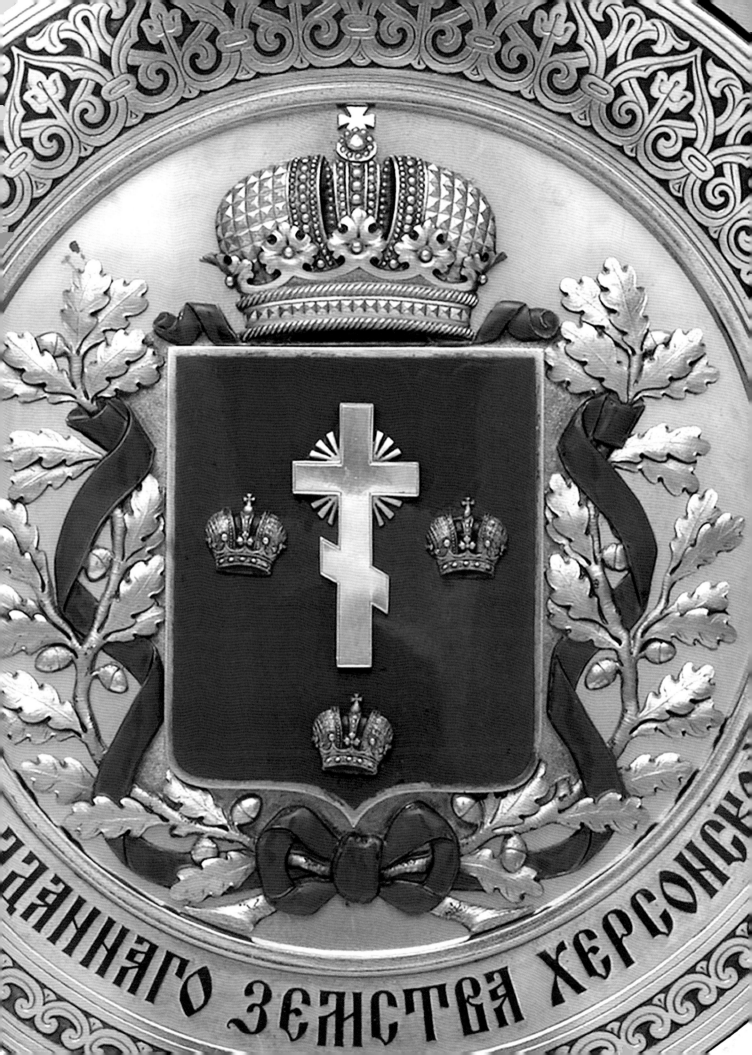

160

盐皿或碗

1899 ~ 1908 年
法贝热公司
工匠大师：埃里克·科林
水晶、黄金、钻石、红宝石
高 34.925 毫米　宽 53.975 毫米
厚 44.45 毫米
弗吉尼亚美术馆藏
莉莲·托马斯·普拉特遗赠

160

Salt Cellar or Bowl

1899–1908
Fabergé firm
Workmaster: Erik Kollin
Rock crystal, gold, diamonds, rubies
1 3/8" H × 2 1/8" W × 1 3/4" D
Virginia Museum of Fine Arts
Bequest of Lillian Thomas Pratt

161

盘

1899 ~ 1908 年
费多尔·吕克特
银镀金、珐琅
高 17.463 毫米　直径 184.15 毫米
弗吉尼亚美术馆藏
杰罗姆和丽塔·甘斯俄罗斯珐琅藏品

161

Plate

1899–1908
Fedor Rückert
Silver gilt, enamel
11/16" H × 7 1/4" Dia.
Virginia Museum of Fine Arts
Jerome and Rita Gans Collection of Russian Enamel

162

果子露杯、托盘和勺

1908～1917年

法贝热公司

工匠大师：伊凡·赫列勃尼科夫

银镀金、珐琅

杯子：高174.625毫米　直径114.3毫米

托盘：高19.050毫米　直径190.5毫米

勺子：高139.7毫米　宽38.1毫米　厚19.05毫米

弗吉尼亚美术馆藏

莎拉·麦高恩捐赠

162

Sherbet Cup, Saucer, and Spoon

1908–1917

Fabergé firm

Workmaster: Ivan Khlebnikov

Silver gilt, enamels

Cup: 6 7/8" H × 4 1/2" Dia.

Saucer: 3/4" H × 7 1/2" Dia.

Spoon: 5 1/2" H × 1 1/2" W × 3/4" D

Virginia Museum of Fine Arts

Gift of Sarah McGowan

163

咖啡具三件套

1908 ~ 1917 年
费多尔·吕克特
银镀金、珐琅、西伯利亚硬岩
高 206.375 毫米　宽 177.8 毫米　厚 120.65 毫米
高 130.175 毫米　宽 127 毫米　厚 101.6 毫米
高 177.8 毫米　宽 152.4 毫米　厚 114.3 毫米
弗吉尼亚美术馆藏
杰罗姆和丽塔·甘斯俄罗斯珐琅藏品

163

Three-Piece Coffee Set

1908–1917
Fedor Rückert
Silver gilt, enamel, Siberian hardstones
8 1/8" H × 7" W × 4 3/4" D
5 1/8" H × 5" W × 4" D
7" H × 6" W × 4 1/2" D
Virginia Museum of Fine Arts
Jerome and Rita Gans Collection of Russian Enamel

164

小咖啡勺

1908～1917 年
法贝热公司
银、银镀金、珐琅
长 117.475 毫米　宽 22.225 毫米
长 115.888 毫米　宽 28.575 毫米
长 142.875 毫米　宽 34.925 毫米
弗吉尼亚美术馆藏
克利福德·马特洛克先生捐赠，
以纪念亡妻尼娜·肖雷平·马特洛克

164

Demitasse Spoons

1908–1917
Fabergé firm
Silver, silver gilt, enamel
4 5/8" L × 7/8" W
4 9/16" L × 1 1/8" W
5 5/8" L × 1 3/8" W
Virginia Museum of Fine Arts
Gift of Mr. Clifford C. Matlock in memory
of his late wife, Nina Syolypin(a) Matlock

165

公用匙

1908 ~ 1917 年
费多尔·吕克特
银镀金、珐琅
高 209.55 毫米　宽 57.15 毫米
厚 31.75 毫米
弗吉尼亚美术馆藏
杰罗姆和丽塔·甘斯俄罗斯珐琅藏品

165

Serving Spoon

1908–1917
Fedor Rückert
Silver gilt, enamel
8 1/4" H × 2 1/4" W × 1 1/4" D
Virginia Museum of Fine Arts
Jerome and Rita Gans Collection
of Russian Enamel

166
第一次世界大战纪念碟

1914 年
法贝热公司
软玉、银镀金
高 44.45 毫米　直径 203.2 毫米
弗吉尼亚美术馆藏
福尔曼·赫布捐赠

166
World War I Dish

1914
Fabergé firm
Nephrite, silver gilt
1 3/4" H × 8" Dia.
Virginia Museum of Fine Arts
Gift of Furman Hebb

167

糖果盒

1899 年以前
法贝热公司
工匠大师：米哈伊尔·佩尔欣
软玉、黄金、银、红宝石、钻石
高 31.75 毫米　直径 41.275 毫米
弗吉尼亚美术馆藏
莉莲·托马斯·普拉特遗赠

167

Bonbonnière

Before 1899
Fabergé firm
Workmaster: Mikhail Perkhin
Nephrite, gold, silver, rubies, diamonds
1 1/4" H × 1 5/8" Dia.
Virginia Museum of Fine Arts
Bequest of Lillian Thomas Pratt

168

糖果盒

1899 ~ 1908 年
法贝热公司
工匠大师：埃里克·科林
水晶、黄金、钻石、红宝石
高 25.4 毫米　宽 47.625 毫米　厚 38.1 毫米
弗吉尼亚美术馆藏
亨利·斯潘塞博士夫妇捐赠

168

Bonbonnière

1899–1908
Fabergé firm
Workmaster: Erik Kollin
Rock crystal, gold, diamonds, ruby
1" H × 1 7/8" W × 1 1/2" D
Virginia Museum of Fine Arts
Gift of Dr. and Mrs. Henry S. Spencer

169

糖果盒

1899～1908年
法贝热公司
黄金、珐琅、钻石
高 19.05 毫米　直径 53.975 毫米
弗吉尼亚美术馆藏
莉莲·托马斯·普拉特遗赠

169

Bonbonnière

1899–1908
Fabergé firm
Gold, enamel, diamonds
3/4" H × 2 1/8" Dia.
Virginia Museum of Fine Arts
Bequest of Lillian Thomas Pratt

170

糖果盒

1899～1908年
法贝热公司
银镀金、珐琅、钻石、红宝石、珍珠
高 22.225 毫米　直径 44.45 毫米
弗吉尼亚美术馆藏
莉莲·托马斯·普拉特遗赠

170

Bonbonnière

1899–1908
Fabergé firm
Silver gilt, enamel, diamonds, rubies, pearls
7/8" H × 1 3/4" Dia.
Virginia Museum of Fine Arts
Bequest of Lillian Thomas Pratt

171

香烟盒

1899 年以前
法贝热公司
工匠大师：奥古斯特·霍尔明
黄金、玛瑙、钻石
高 31.75 毫米　长 88.9 毫米　宽 41.275 毫米
弗吉尼亚美术馆藏
艾尔萨·梅隆·布鲁斯家族捐赠

171

Cigarette Case

Before 1899
Fabergé firm
Workmaster: August Hollming
Gold, agate, diamonds
1 1/4" H × 3 1/2" L × 1 5/8" W
Virginia Museum of Fine Arts
Gift of the Estate of Ailsa Mellon Bruce

172

香烟盒

1899 年以前
法贝热公司
工匠大师：米哈伊尔·佩尔欣
银镀金、珐琅、红玉髓、钻石
高 28.575 毫米　长 85.725 毫米　宽 41.275 毫米
弗吉尼亚美术馆藏
艾尔萨·梅隆·布鲁斯家族捐赠

172

Cigarette Case

Before 1899
Fabergé firm
Workmaster: Mikhail Perkhin
Silver gilt, enamel, carnelian, diamonds
1 1/8" H × 3 3/8" L × 1 5/8" W
Virginia Museum of Fine Arts
Gift of the Estate of Ailsa Mellon Bruce

173

香烟盒

1899～1908年
法贝热公司
工匠大师：米哈伊尔·佩尔欣
银镀金、珐琅、红宝石
高107.95毫米　宽66.675毫米
厚12.7毫米
弗吉尼亚美术馆藏
福尔曼·赫布捐赠

173

Cigarette Case

1899–1908
Fabergé firm
Workmaster: Mikhail Perkhin
Silver gilt, enamel, ruby
4 1/4" H × 2 5/8" W × 1/2" D
Virginia Museum of Fine Arts
Gift of Furman Hebb

174

香烟盒

1899～1908 年
法贝热公司
黄金、红宝石
高 98.425 毫米　宽 79.375 毫米
厚 12.7 毫米
弗吉尼亚美术馆藏
莉莲·托马斯·普拉特遗赠

174

Cigarette Case

1899–1908
Fabergé firm
Gold, rubies
3 7/8" H × 3 1/8" W × 1/2" D
Virginia Museum of Fine Arts
Bequest of Lillian Thomas Pratt

175

香烟盒

1899～1908 年
法贝热公司
工匠大师：亨瑞克·魏格斯特姆
黄金、珐琅、钻石
长 85.725 毫米　直径 33.338 毫米
弗吉尼亚美术馆藏
艾尔萨·梅隆·布鲁斯家族捐赠

175

Cigarette Case

1899–1908
Fabergé firm
Workmaster: Henrik Wigström
Gold, enamel, diamonds
3 3/8" L × 1 5/16" Dia.
Virginia Museum of Fine Arts
Gift of the Estate of Ailsa Mellon Bruce

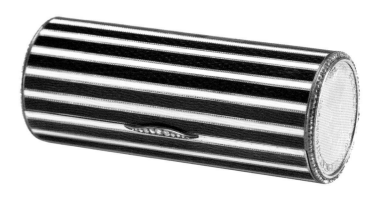

176

图饰珐琅纸牌盒

1899 年以前
伊凡·萨尔特科夫
银镀金、珐琅
高 107.95 毫米　宽 69.85 毫米
厚 50.8 毫米
弗吉尼亚美术馆藏
杰罗姆和丽塔·甘斯俄罗斯珐琅藏品

176

Pictorial Enamel Cardcase

Before 1899
Ivan Saltykov
Silver gilt, enamel
4 1/4" H × 2 3/4" W × 2" D
Virginia Museum of Fine Arts
Jerome and Rita Gans Collection
of Russian Enamel

177

象盒

1899 年以前
法贝热公司
工匠大师：米哈伊尔·佩尔欣
软玉、猛犸象牙、黄金、银、红宝石、钻石
高 82.55 毫米　直径 95.25 毫米
弗吉尼亚美术馆藏
莉莲·托马斯·普拉特遗赠

178

皇家礼品盒

1899～1908 年
法贝热公司
工匠大师：米哈伊尔·佩尔欣
软玉、黄金、银、钻石
高 41.276 毫米　直径 79.375 毫米
弗吉尼亚美术馆藏
莉莲·托马斯·普拉特遗赠

177

Elephant Box

Before 1899
Fabergé firm
Workmaster: Mikhail Perkhin
Nephrite, Mammoth ivory, gold, silver, rubies, diamonds
3 1/4" H × 3 3/4" Dia.
Virginia Museum of Fine Arts
Bequest of Lillian Thomas Pratt

178

Imperial Presentation Box

1899–1908
Fabergé firm
Workmaster: Mikhail Perkhin
Nephrite, gold, silver, diamonds
1 5/8" H × 3 1/8" Dia.
Virginia Museum of Fine Arts
Bequest of Lillian Thomas Pratt

179

阿拉伯文小盒

1899～1908 年
法贝热公司
工匠大师：米哈伊尔·佩尔欣
黄金、银、珐琅、红玉髓、钻石
高 22.225 毫米　宽 38.1 毫米
厚 28.575 毫米
弗吉尼亚美术馆藏
莉莲·托马斯·普拉特遗赠

179

Box with Arabic Inscription

1899–1908
Fabergé firm
Workmaster: Mikhail Perkhin
Gold, silver, enamel, carnelian, diamonds
7/8" H × 1 1/2" W × 1 1/8" D
Virginia Museum of Fine Arts
Bequest of Lillian Thomas Pratt

180

盒子

1899～1908 年
法贝热公司
工匠大师：亨瑞克·魏格斯特姆
软玉、黄金、钻石、珐琅
高 12.7 毫米　长 69.85 毫米
宽 25.4 毫米
弗吉尼亚美术馆藏
艾尔萨·梅隆·布鲁斯家族捐赠

180

Box

1899–1908
Fabergé firm
Workmaster: Henrik Wigström
Nephrite, gold, diamonds, enamel
1/2" H × 2 3/4" L × 1" W
Virginia Museum of Fine Arts
Gift of the Estate of Ailsa Mellon Bruce

181

盒子

1899 ~ 1908 年
法贝热公司
工匠大师：亨瑞克・魏格斯特姆
软玉、黄金、银镀金、钻石
高 25.4 毫米　长 63.5 毫米
宽 34.925 毫米
弗吉尼亚美术馆藏
艾尔萨・梅隆・布鲁斯家族捐赠

181

Box

1899–1908
Fabergé firm
Workmaster: Henrik Wigström
Nephrite, gold, silver gilt, diamonds
1" H × 2 1/2" L × 1 3/8" W
Virginia Museum of Fine Arts
Gift of the Estate of Ailsa Mellon Bruce

182

盒子

1908 ~ 1917 年
费多尔·吕克特
银镀金、珐琅
高 57.15 毫米　宽 142.875 毫米
厚 117.475 毫米
弗吉尼亚美术馆藏
杰罗姆和丽塔·甘斯俄罗斯珐琅藏品

182

Box

1908–1917
Fedor Rückert
Silver gilt, enamel
2 1/4" H × 5 5/8" W × 4 5/8" D
Virginia Museum of Fine Arts
Jerome and Rita Gans Collection
of Russian Enamel

183

盒子

19～20 世纪，1947 年以前
未知
法贝热仿制品
黄金、银、珐琅、钻石、红宝石、珍珠
高 42.863 毫米　直径 92.075 毫米
弗吉尼亚美术馆藏
莉莲·托马斯·普拉特遗赠

作为普拉特藏品中最为重要的作品之一，这件沉甸甸的皇家礼盒虽然是被当作法贝热真品购得，但有趣的是，它实际上却是一件赝品。礼盒的制造者显然不清楚法贝热工作室的标记系统。他给礼盒打上工匠大师米哈伊尔·佩尔欣（主要工作地点在圣彼得堡）的姓名首字母，法贝热的西里尔字母"K. 法贝热（K. Fabergé）"皇家认证标记，又加上一只沙俄皇室象征双头鹰。然而，法贝热的皇家认证标记只用在产自莫斯科的作品上。而且，礼盒内里顶盖饰有珐琅，而盒壁与底部却无本应当有的珐琅装饰层，说明这只礼盒在风格上更接近法贝热的竞争对手。

183

Box

19th–20th Century, before 1947
Unknown
Fabergé forgery
Gold, silver, enamel, diamonds, rubies, pearls
1 11/16" H × 3 5/8" Dia.
Virginia Museum of Fine Arts
Bequest of Lillian Thomas Pratt

One of the most prominent objects in the Pratt collection, this heavy, imperial presentation box that sold as a Fabergé original, is an interesting forgery. Its maker was unaware of the hallmarking system used in the Fabergé workshops. The box is struck with the initials of workmaster Mikhail Perkin (who mainly worked in St. Petersburg) as well as Fabergé's imperial warrant mark, "K. Fabergé" in Cyrillic, and is set with the Russian double-headed eagle. However, this imperial mark was used only for objects produced in Moscow. In addition, the enameled interior cover and the lack of the usual enamel coating on the sides and base reveal that this box is stylistically closer to works by Fabergé's competitors.

184

珐琅图饰盒

1908～1917年
法贝热公司
银镀金、珐琅
高 22.225 毫米　直径 53.975 毫米
弗吉尼亚美术馆藏
杰罗姆和丽塔·甘斯俄罗斯珐琅藏品

这件圆形珐琅图饰盒无疑是费奥多尔·吕克特工坊制作的，盒盖和盒外侧均为老式俄罗斯风格装饰。盒盖上仿的是维克特·瓦斯涅佐夫的名画《伊凡王子骑灰狼》，这个故事也是斯特拉文斯基的芭蕾舞剧《火鸟》的灵感源泉。

184

Enamel Pictorial Box

1908–1917
Fabergé firm
Silver gilt, enamel
7/8" H × 2 1/8" Dia.
Virginia Museum of Fine Arts
Jerome and Rita Gans Collection of Russian Enamel

The cover and sides of this circular box, which is certainly from the workshop of Fedor Rückert, are decorated in the Old Russian style. The cover is painted with a copy of Viktor Vasnetsov's celebrated painting Ivan Tsarevich Rides the Grey Wolf, the tale that inspired Stravinsky's The Firebird.

185

鱼子酱碗

1896～1908 年
法贝热公司
银
高 184.15 毫米　长 577.85 毫米　厚 177.8 毫米
弗吉尼亚美术馆藏
杰拉尔德·德希尔瓦先生藏品

这只鱼子酱碗展现了法贝热工匠在处理白银等贵金属方面的精湛技艺。选择鲟鱼作为这只鱼子酱碗的造型体现了法贝热的主顾奢华的生活方式——以俄罗斯鲟鱼鱼卵为原料制作的鱼子酱品质最佳，因此鲟鱼成为俄罗斯最昂贵的鱼类。它们也经常出现在俄罗斯传说故事之中，如有关商人之子萨特阔的传说：萨特阔拜访海神时，发现了一对守卫海底宫殿的鲟鱼。

185

Caviar Bowl

1896–1908
Fabergé firm
Silver
7 1/4" H × 22 3/4" L × 7" D
Virginia Museum of Fine Arts
Collection of Mr. Gerald M. de Sylvar

This caviar bowl shows the skill of Fabergé's craftsmen in manipulating precious metals such as silver. The choice of a sturgeon for the shape of the bowl reflects the opulent lifestyle of Fabergé's patrons—Russian sturgeons produce superior caviar, making them the most valuable of Russian fish. They also figure in Russian tales, including the story of Sadko the Merchant's Son; Sadko visits the tsar of the sea and finds a pair of sturgeons guarding the underwater palace.

186

鼻烟壶

1899～1908 年
法贝热公司
石英或长石、黄金、钻石、红宝石、
祖母绿、蓝宝石、珍珠、珐琅
高 25.4 毫米　长 53.975 毫米　宽 44.45 毫米
弗吉尼亚美术馆藏
莉莲・托马斯・普拉特遗赠

186

Snuffbox

1899–1908
Fabergé firm
Quartz or feldspar, gold, diamond,
ruby, emerald, sapphire, pearl, enamel
1" H × 2 1/8" L × 1 3/4" W
Virginia Museum of Fine Arts
Bequest of Lillian Thomas Pratt

187

第一次世界大战纪念烟灰缸

1914 年
法贝热公司
铜
高 34.925 毫米　直径 107.95 毫米
弗吉尼亚美术馆藏
莉莲・托马斯・普拉特遗赠

187

World War I Ashtray

1914
Fabergé firm
Copper
1 3/8" H × 4 1/4" Dia.
Virginia Museum of Fine Arts
Bequest of Lillian Thomas Pratt

188

墨水池

1899 ~ 1908 年
法贝热公司
工匠大师：费奥多尔·阿法纳西耶夫
软玉、黄金、银镀金、珐琅、水晶
高 98.425 毫米 直径 76.2 毫米
弗吉尼亚美术馆藏
莉莲·托马斯·普拉特遗赠

188

Inkwell

1899–1908
Fabergé firm
Workmaster: Fedor Afanas'ev
Nephrite, gold, silver gilt, enamel, rock crystal
3 7/8" H × 3" Dia.
Virginia Museum of Fine Arts
Bequest of Lillian Thomas Pratt

189

花瓶

1899 ~ 1908 年
法贝热公司
工匠大师：安蒂·内瓦莱南
银镀金、珐琅
高 92.075 毫米　直径 66.675 毫米
弗吉尼亚美术馆藏
莉莲·托马斯·普拉特遗赠

  这只饰有月桂垂花饰的花瓶属于新古典主义风格。可能是用来架香烟的，瓶身的红色珐琅色泽浓厚，与银镀金的装饰图案相映成趣。法贝热深受 18 世纪后期的法国金匠和珠宝商的影响，他们也正是当时风格最超前的金属匠艺人。

189

Vase

1899–1908
Fabergé firm
Workmaster: Antti Nevalainen
Silver gilt, enamel
3 5/8" H × 2 5/8 " Dia.
Virginia Museum of Fine Arts
Bequest of Lillian Thomas Pratt

  This vase, ornamented with laurel swags, is in the neoclassical style. Probably intended to hold cigarettes, the container's red enamel provides a rich background for the silver gilt ornamentation. Fabergé was strongly influenced by French goldsmiths and jewelers of the late eighteenth century, who were the most stylistically advanced metalsmiths of their era.

190

硬币花瓶

1908 ~ 1917 年
法贝热公司
工匠大师：韦克娃
银
高 95.25 毫米　直径 120.65 毫米
弗吉尼亚美术馆藏
莉莲·托马斯·普拉特遗赠

190

Coin Vase

1908–1917
Fabergé firm
Workmaster: J. Wakeva
Silver
3 3/4" H × 4 3/4" Dia.
Virginia Museum of Fine Arts
Bequest of Lillian Thomas Pratt

191

书签

1899 年以前
法贝热公司
工匠大师：米哈伊尔·佩尔欣
鲍文玉（硬绿蛇纹石）、红宝石、钻石、黄金
高 44.45 毫米　宽 31.75 毫米
厚 9.525 毫米
弗吉尼亚美术馆藏
莉莲·托马斯·普拉特遗赠

191

Bookmark

Before 1899
Fabergé firm
Workmaster: Mikhail Perkhin
Bowenite, rubies, diamonds, gold
1 3/4" H × 1 1/4" W × 3/8" D
Virginia Museum of Fine Arts
Bequest of Lillian Thomas Pratt

192

地球仪

1899 年以前
法贝热公司
工匠大师：埃里克·科林
水晶、黄金、指南针
高 133.35 毫米　直径 82.55 毫米
弗吉尼亚美术馆藏
莉莲·托马斯·普拉特遗赠

192

Terrestrial Globe

Before 1899
Fabergé firm
Workmaster: Erik Kollin
Rock crystal, gold, compass
5 1/4" H × 3 1/4" Dia.
Virginia Museum of Fine Arts
Bequest of Lillian Thomas Pratt

193

项坠

1899～1903 年
法贝热公司
工匠大师：米哈伊尔·佩尔欣
黄金、银镀金、钻石、红宝石、珐琅
高 41.275 毫米　宽 28.575 毫米　厚 9.525 毫米
弗吉尼亚美术馆藏
莉莲·托马斯·普拉特遗赠

193

Locket

1899–1903
Fabergé firm
Workmaster: Mikhail Perkhin
Gold, silver gilt, diamonds,
rubies, enamel
1 5/8" H × 1 1/8" W × 3/8" D
Virginia Museum of Fine Arts
Bequest of Lillian Thomas Pratt

194

沙皇孤儿院徽章

（带有皇后亚历山德拉·费奥多罗芙娜的花押）

1907 年
法贝热公司
工匠大师：阿尔佛雷德·蒂勒曼
黄金、银、珐琅
高 34.925 毫米　宽 25.4 毫米　厚 6.35 毫米
弗吉尼亚美术馆藏
福尔曼·赫布捐赠

194

Badge of the Imperial Orphanage
with the Cypher of
Tsaritsa Alexandra Feodorovna

1907
Fabergé firm
Workmaster: Alfred Thielemann
Gold, silver, enamel
1 3/8" H × 1" W × 1/4" D
Virginia Museum of Fine Arts
Gift of Furman Hebb

195

万年历

1908～1917年
法贝热公司
工匠大师：亨瑞克·魏格斯特姆
软玉、黄金、银镀金、珐琅、玛瑙
高 88.9 毫米　宽 98.425 毫米
厚 15.875 毫米
弗吉尼亚美术馆藏
莉莲·托马斯·普拉特遗赠

195

Perpetual Calendar

1908–1917
Fabergé firm
Workmaster: Henrik Wigström
Nephrite, gold, silver gilt, enamel, agate
3 1/2" H × 3 7/8" W × 5/8" D
Virginia Museum of Fine Arts
Bequest of Lillian Thomas Pratt

196

伞柄

1899 年以前
法贝热公司
工匠大师：米哈伊尔·佩尔欣
鲍文玉（硬绿蛇纹石）、黄金、银、钻石、红宝石、珐琅
高 73.025 毫米　宽 53.975 毫米　厚 25.4 毫米
弗吉尼亚美术馆藏
莉莲·托马斯·普拉特遗赠

169

Parasol Handle

Before 1899
Fabergé firm
Workmaster: Mikhail Perkhin
Bowenite, gold, silver, diamonds, rubies, enamel
2 7/8" H × 2 1/8" W × 1" D
Virginia Museum of Fine Arts
Bequest of Lillian Thomas Pratt

197

伞柄

1899 年以前
法贝热公司
工匠大师：埃里克·科林
软玉、黄金、钻石、红宝石
高 57.15 毫米　直径 38.1 毫米
弗吉尼亚美术馆藏
莉莲·托马斯·普拉特遗赠

197

Parasol Handle

Before 1899
Fabergé firm
Workmaster: Erik Kollin
Nephrite, gold, diamonds, rubies
2 1/4" H × 1 1/2" Dia.
Virginia Museum of Fine Arts
Bequest of Lillian Thomas Pratt

198

伞柄

1899 年以前
法贝热公司
黄金、珐琅、黄玉
高 60.325 毫米　直径 28.575 毫米
弗吉尼亚美术馆藏
莉莲・托马斯・普拉特遗赠

198

Parasol Handle

Before 1899
Fabergé firm
Gold, enamel, topaz
2 3/8" H × 1 1/8" Dia.
Virginia Museum of Fine Arts
Bequest of Lillian Thomas Pratt

199

伞柄

1899 年以前
法贝热公司
工匠大师：米哈伊尔・佩尔欣
鲍文玉（硬绿蛇纹石）、黄金、珐琅
高 120.65 毫米　直径 19.05 毫米
弗吉尼亚美术馆藏
莉莲・托马斯・普拉特遗赠

199

Parasol Handle

Before 1899
Fabergé firm
Workmaster: Mikhail Perkhin
Nephrite, gold, enamel
4 3/4" H × 3/4" Dia.
Virginia Museum of Fine Arts
Bequest of Lillian Thomas Pratt

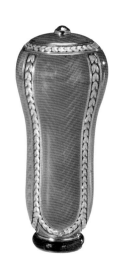

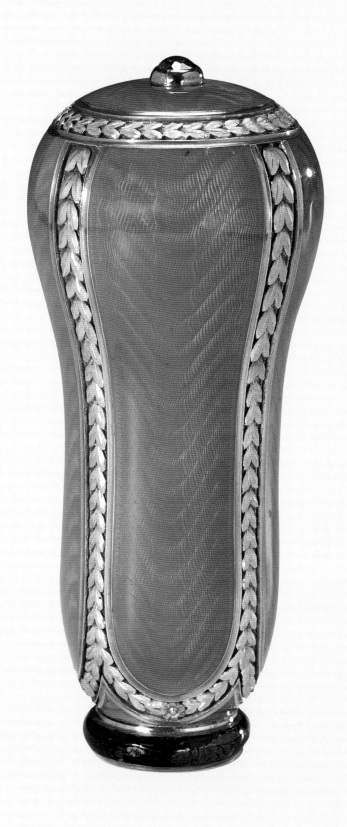

200

伞柄

1899 年以前
法贝热公司
工匠大师：米哈伊尔·佩尔欣
水晶、黄金、银、珐琅、钻石
高 76.2 毫米　直径 22.225 毫米
弗吉尼亚美术馆藏
莉莲·托马斯·普拉特遗赠

200

Parasol Handle

Before 1899
Fabergé firm
Workmaster: Mikhail Perkhin
Rock crystal, gold, silver, enamel, diamonds
3" H × 7/8" Dia.
Virginia Museum of Fine Arts
Bequest of Lillian Thomas Pratt

201

伞柄

1899 ～ 1908 年
法贝热公司
东陵玉、黄金、珐琅、钻石
高 63.5 毫米　长 79.375 毫米　宽 15.875 毫米
弗吉尼亚美术馆藏
莉莲·托马斯·普拉特遗赠

201

Parasol Handle

1899–1908
Fabergé firm
Aventurine, gold, enamel, diamonds
2 1/2" H × 3 1/8" L × 5/8" W
Virginia Museum of Fine Arts
Bequest of Lillian Thomas Pratt

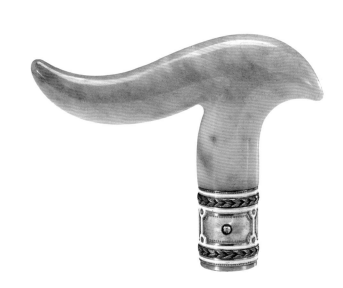

202

伞柄

1899～1908 年
法贝热公司
黄金、银、钻石
高 60.325 毫米　宽 41.275 毫米　厚 15.875 毫米
弗吉尼亚美术馆藏
莉莲·托马斯·普拉特遗赠

202

Parasol Handle

1899–1908
Fabergé firm
Gold, silver, diamonds
2 3/8" H × 1 5/8" W × 5/8" D
Virginia Museum of Fine Arts
Bequest of Lillian Thomas Pratt

203

伞柄

1899～1908 年
法贝热公司
工匠大师：亨瑞克·魏格斯特姆
鲍文玉（硬绿蛇纹石）、黄金、红宝石
高 95.25 毫米　直径 50.8 毫米
弗吉尼亚美术馆藏
莉莲·托马斯·普拉特遗赠

203

Parasol Handle

1899–1908
Fabergé firm
Workmaster: Henrik Wigström
Bowenite, gold, ruby
3 3/4" H × 2" Dia.
Virginia Museum of Fine Arts
Bequest of Lillian Thomas Pratt

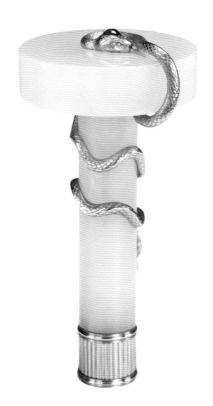

204

伞柄

1899 ~ 1908 年
法贝热公司
工匠大师：亨瑞克·魏格斯特姆
黄金、珐琅、钻石
高 73.025 毫米　直径 22.225 毫米
弗吉尼亚美术馆藏
莉莲·托马斯·普拉特遗赠

204

Parasol Handle

1899–1908
Fabergé firm
Workmaster: Henrik Wigström
Gold, enamel, diamond
2 7/8" H × 7/8" Dia.
Virginia Museum of Fine Arts
Bequest of Lillian Thomas Pratt

205

伞柄

1899 ~ 1908 年
法贝热公司
工匠大师：米哈伊尔·佩尔欣
水晶、黄金、银、银镀金、红宝石
高 85.725 毫米　直径 25.4 毫米
弗吉尼亚美术馆藏
莉莲·托马斯·普拉特遗赠

205

Parasol Handle

1899–1908
Fabergé firm
Workmaster: Mikhail Perkhin
Rock crystal, gold, silver, silver gilt, rubies
3 3/8" H × 1" Dia.
Virginia Museum of Fine Arts
Bequest of Lillian Thomas Pratt

206

伞柄

约 1900 年
法贝热公司
工匠大师：米哈伊尔·佩尔欣
鲍文玉（硬绿蛇纹石）、黄金、银、钻石、珐琅
高 57.15 毫米　长 76.2 毫米　宽 19.05 毫米
弗吉尼亚美术馆藏
莉莲·托马斯·普拉特遗赠

206

Parasol Handle

ca. 1900
Fabergé firm
Workmaster: Mikhail Perkhin
Bowenite, gold, silver, diamonds, enamel
2 1/4" H × 3" L × 3/4" W
Virginia Museum of Fine Arts
Bequest of Lillian Thomas Pratt

207

伞柄

约 1900 年
法贝热公司
工匠大师：米哈伊尔·佩尔欣
软玉、黄金、钻石
高 85.725 毫米　宽 36.513 毫米　厚 31.75 毫米
弗吉尼亚美术馆藏
莉莲·托马斯·普拉特遗赠

207

Parasol Handle

ca. 1900
Fabergé firm
Workmaster: Mikhail Perkhin
Nephrite, gold, diamonds
3 3/8" H × 1 7/16" W × 1 1/4" D
Virginia Museum of Fine Arts
Bequest of Lillian Thomas Pratt

208 伞柄 约 1900 年 法贝热公司 工匠大师：亨瑞克·魏格斯特姆 软玉、黄金、银镀金、珐琅、玉髓 高 50.8 毫米　长 85.725 毫米　宽 19.05 毫米 弗吉尼亚美术馆藏 莉莲·托马斯·普拉特遗赠	209 伞柄 约 1900 年 法贝热公司 工匠大师：米哈伊尔·佩尔欣 鲍文玉（硬绿蛇纹石）、黄金、银、钻石、珍珠、珐琅 高 63.5 毫米　宽 38.1 毫米　厚 19.05 毫米 弗吉尼亚美术馆藏 莉莲·托马斯·普拉特遗赠

208

Parasol Handle

ca. 1900

Fabergé firm

Workmaster: Henrik Wigström

Nephrite, gold, silver gilt, enamel, chalcedony

2" H × 3 3/8" L × 3/4" W

Virginia Museum of Fine Arts

Bequest of Lillian Thomas Pratt

209

Parasol Handle

ca. 1900

Fabergé firm

Workmaster: Mikhail Perkhin

Bowenite, gold, silver, diamonds, pearls, enamel

2 1/2" H × 1 1/2" W × 3/4" D

Virginia Museum of Fine Arts

Bequest of Lillian Thomas Pratt

210

伞柄

约 1900 年
法贝热公司
鲍文玉（硬绿蛇纹石）、黄金、珐琅、珍珠
高 76.2 毫米　直径 25.4 毫米
弗吉尼亚美术馆藏
莉莲·托马斯·普拉特遗赠

210

Parasol Handle

ca. 1900
Fabergé firm
Bowenite, gold, enamel, pearls
3" H × 1" Dia.
Virginia Museum of Fine Arts
Bequest of Lillian Thomas Pratt

211

伞柄礼盒

约 1900 年
俄罗斯
枫木、天鹅绒、绸缎、黄铜
高 31.75 毫米　宽 95.25 毫米　厚 50.8 毫米
弗吉尼亚美术馆藏
莉莲·托马斯·普拉特遗赠

211

Presentation Box
for Parasol Handle

ca. 1900
Russian
Maple, velvet, satin, brass
1 1/4" H × 3 3/4" W × 2" D
Virginia Museum of Fine Arts
Bequest of Lillian Thomas Pratt

212

伞柄

1908 ~ 1917 年
法贝热公司
水晶、黄金、珐琅、珍珠
高 88.9 毫米　宽 44.45 毫米　厚 15.875 毫米
弗吉尼亚美术馆藏
莉莲·托马斯·普拉特遗赠

212

Parasol Handle

1908–1917
Fabergé firm
Rock crystal, gold, enamel, pearls
3 1/2" H × 1 3/4" W × 5/8" D
Virginia Museum of Fine Arts
Bequest of Lillian Thomas Pratt

213

手杖柄

1893 年以前
法贝热公司
工匠大师：米哈伊尔·佩尔欣
鲍文玉（硬绿蛇纹石）、黄金、银、珐琅、钻石
高 76.2 毫米　宽 44.45 毫米　厚 25.4 毫米
弗吉尼亚美术馆藏
莉莲·托马斯·普拉特遗赠

213

Cane Handle

Before 1893
Fabergé firm
Workmaster: Mikhail Perkhin
Bowenite, gold, silver, enamel, diamonds
3" H × 1 3/4" W × 1" D
Virginia Museum of Fine Arts
Bequest of Lillian Thomas Pratt

259

214

手杖柄

1899 年以前
法贝热公司
工匠大师：米哈伊尔·佩尔欣
黄金、银、珐琅、钻石
高 69.85 毫米　直径 41.275 毫米
弗吉尼亚美术馆藏
莉莲·托马斯·普拉特遗赠

215

手杖柄

1899 年以前
法贝热公司
工匠大师：米哈伊尔·佩尔欣
鲍文玉（硬绿蛇纹石）、黄金、钻石
高 66.675 毫米　宽 34.925 毫米　厚 28.575 毫米
弗吉尼亚美术馆藏
莉莲·托马斯·普拉特遗赠

214

Cane Handle

Before 1899
Fabergé firm
Workmaster: Mikhail Perkhin
Gold, silver, enamel, diamonds
2 3/4" H × 1 5/8" Dia.
Virginia Museum of Fine Arts
Bequest of Lillian Thomas Pratt

215

Cane Handle

Before 1899
Fabergé firm
Workmaster: Mikhail Perkhin
Bowenite, gold, diamonds
2 5/8" H × 1 3/8" W × 1 1/8" D
Virginia Museum of Fine Arts
Bequest of Lillian Thomas Pratt

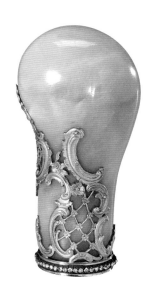

216

手杖柄

1899 年以前
法贝热公司
工匠大师：米哈伊尔·佩尔欣
鲍文玉（硬绿蛇纹石）、黄金、银、珐琅、钻石、珍珠
高 63.5 毫米　直径 38.1 毫米
弗吉尼亚美术馆藏
莉莲·托马斯·普拉特遗赠

216

Cane Handle

Before 1899
Fabergé firm
Workmaster: Mikhail Perkhin
Bowenite, gold, silver, enamel, diamonds, pearls
2 1/2" H × 1 1/2" Dia.
Virginia Museum of Fine Arts
Bequest of Lillian Thomas Pratt

217

手杖柄

1899 年以前
法贝热公司
工匠大师：米哈伊尔·佩尔欣
黄金、银镀金、钻石、珐琅
高 82.55 毫米　宽 34.925 毫米
弗吉尼亚美术馆藏
莉莲·托马斯·普拉特遗赠

217

Cane Handle

Before 1899
Fabergé firm
Workmaster: Mikhail Perkhin
Gold, silver gilt, diamonds, enamel
3 1/4" H × 1 3/8" W
Virginia Museum of Fine Arts
Bequest of Lillian Thomas Pratt

218

手杖柄

1899 年以前
法贝热公司
工匠大师：米哈伊尔·佩尔欣
水晶、黄金、银、珐琅、钻石、祖母绿
高 60.325 毫米　长 92.075 毫米　宽 22.225 毫米
弗吉尼亚美术馆藏
莉莲·托马斯·普拉特遗赠

218

Cane Handle

Before 1899
Fabergé firm
Workmaster: Mikhail Perkhin
Rock crystal, gold, silver, enamel, diamonds, emeralds
2 3/8" H × 3 5/8" L × 7/8" W
Virginia Museum of Fine Arts
Bequest of Lillian Thomas Pratt

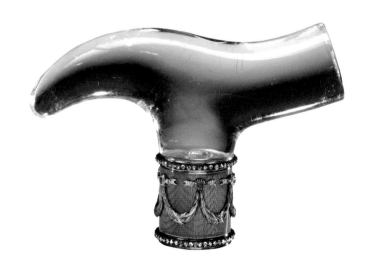

219	220
手杖柄	手杖柄

219

1899～1908年
法贝热公司
水晶、黄金、珐琅、钻石
高44.45毫米　长44.45毫米　宽9.525毫米
弗吉尼亚美术馆藏
莉莲·托马斯·普拉特遗赠

220

约1900年
被认为是法贝热公司作品
玛瑙、黄金、珐琅、红宝石
高85.725毫米　宽41.275毫米　厚38.1毫米
弗吉尼亚美术馆藏
莉莲·托马斯·普拉特遗赠

219

Cane Handle

1899–1908
Fabergé firm
Rock crystal, gold, enamel, diamonds
1 3/4" H × 1 3/4" L × 3/8" W
Virginia Museum of Fine Arts
Bequest of Lillian Thomas Pratt

220

Cane Handle

ca. 1900
Attributed to Fabergé firm
Agate, gold, enamel, rubies
3 3/8" H × 1 5/8" W × 1 1/2" D
Virginia Museum of Fine Arts
Bequest of Lillian Thomas Pratt

信仰的力量
Power of Faith

　　复活节是俄罗斯东正教教历中最重要的节日。复活节的庆祝活动包括礼拜、传统的节日问候以及丰盛的宴请。在俄罗斯东正教的文化中,宗教偶像(如耶稣、圣母、圣徒)对信徒的重要性不言而喻。不惜工本地以珍贵材料装饰这些偶像的圣像也是俄罗斯东正教艺术的一大特色。

Easter is the most important festival of the Russian Orthodox Church. Celebration activities include religious services, traditional festival greetings, and sumptuous feasts. It goes without saying how important the religious icons (e.g. Jesus, Virgin Mary, and saints) are to believers in the Russian Orthodox culture. Decorating the statues of these icons at all costs with precious materials is also a prominent feature of Russian Orthodox art.

221

《十二门徒圣像》

17 世纪
俄罗斯
木板油画、金箔、铜鎏金
高 314.325 毫米　宽 269.875 毫米
弗吉尼亚美术馆藏
莉莲·托马斯·普拉特遗赠

221

Icon of Twelve Saints

17th Century
Russian
Oil and gold leaf on wood, gilt brass
12 3/8" H × 10 5/8" W
Virginia Museum of Fine Arts
Bequest of Lillian Thomas Pratt

222

《艾弗斯卡亚圣母》

19 世纪
俄罗斯
木板油画、银镀金、珐琅、丝绸
高 269.875 毫米　宽 222.25 毫米　厚 49.213 毫米
弗吉尼亚美术馆藏
莉莲·托马斯·普拉特遗赠

　　这幅圣像上表面覆盖有一个银镀金外罩（Oklad），即一种用于保护圣像的金属外壳。金属外壳采用镂空工艺，以显示出圣像的面部、手部和足部。外罩通常制作精美，装饰华丽。如本件文物所示，圣母玛利亚和圣子的长袍采用氧化银工艺，光晕则饰有掐丝珐琅。这幅圣像属于沙皇尼古拉二世之女塔季扬娜女大公。

222

The Iverskaya Mother of God

19th Century
Russian
Oil on panel, silver gilt, enamel, silk
10 5/8" H × 8 3/4" W × 1 15/16" D
Virginia Museum of Fine Arts
Bequest of Lillian Thomas Pratt

　　This icon is covered with a gilded silver *oklad*, a metal overlay that protects an icon. Openings are cut into the metal of an *oklad* so that the faces, hands, and feet (if they are shown) of the holy figures can be seen. *Oklads* are usually finely worked and decorated, as in this example in which the Virgin and Child are adorned with oxidized-silver robes and cloisonné-enamel haloes. This icon belonged to Grand Duchess Tatiana, daughter of Tsar Nicholas II.

《抹大拉的圣玛利亚、奇迹创造者圣尼古拉、圣亚历山大·涅夫斯基大公》

1899 年以前
瓦西莉·费多托夫·伊利英
银、银镀金
高 177.8 毫米　宽 127 毫米　厚 15.875 毫米
弗吉尼亚美术馆藏
莉莲·托马斯·普拉特遗赠

  这尊圣像是送给沙皇亚历山大二世和皇后玛利亚·亚历山德罗芙娜之子皇储尼古拉·亚历山大罗维奇的礼物。尼古拉于 1865 年逝世，其弟弟亚历山大·亚历山大罗维奇继位，即后来的沙皇亚历山大三世。

St. Mary Magdalene, St. Nicholas the Miracle Worker, St. Prince Aleksandr Nevskii

Before 1899
Vassilii Fedotov IL'in
Silver, silver gilt
7" H × 5" W × 5/8" D
Virginia Museum of Fine Arts
Bequest of Lillian Thomas Pratt

  This icon was presented to Tsesarevich Nikolai Aleksandrovich, son of Emperor Alexander II and Empress Maria Aleksandrovnam. Nikolai, who died in 1865, was the older brother of Alexandr Aleksandrovich, who became Tsar Alexander III.

224

《圣亚历山德拉公主与奇迹创造者圣尼古拉》

1899 年以前
法贝热公司
工匠大师：埃里克·科林
木板油画、银、绿松石、石榴石、珍珠
高 114.3 毫米　宽 171.45 毫米　厚 19.05 毫米
弗吉尼亚美术馆藏
莉莲·托马斯·普拉特遗赠

224

Princess St.Alexandra and St.Nicholas the Miracle Worker Diptych

Before 1899
Fabergé firm
Workmaster: Erik Kollin
Oil on panel, silver, turquoises, garnets, pearls
4 1/2" H × 6 3/4" W × 3/4" D
Virginia Museum of Fine Arts
Bequest of Lillian Thomas Pratt

225

《全能的基督》

1899 ~ 1908 年
法贝热公司
木板油画、银、珐琅、珍珠
高 228.6 毫米　宽 180.975 毫米　厚 19.05 毫米
弗吉尼亚美术馆藏
莉莲·托马斯·普拉特遗赠

225

Christ Pantocrator

1899–1908
Fabergé firm
Oil on panel, silver, enamel, pearls
9" H × 7 1/8" W × 3/4" D
Virginia Museum of Fine Arts
Bequest of Lillian Thomas Pratt

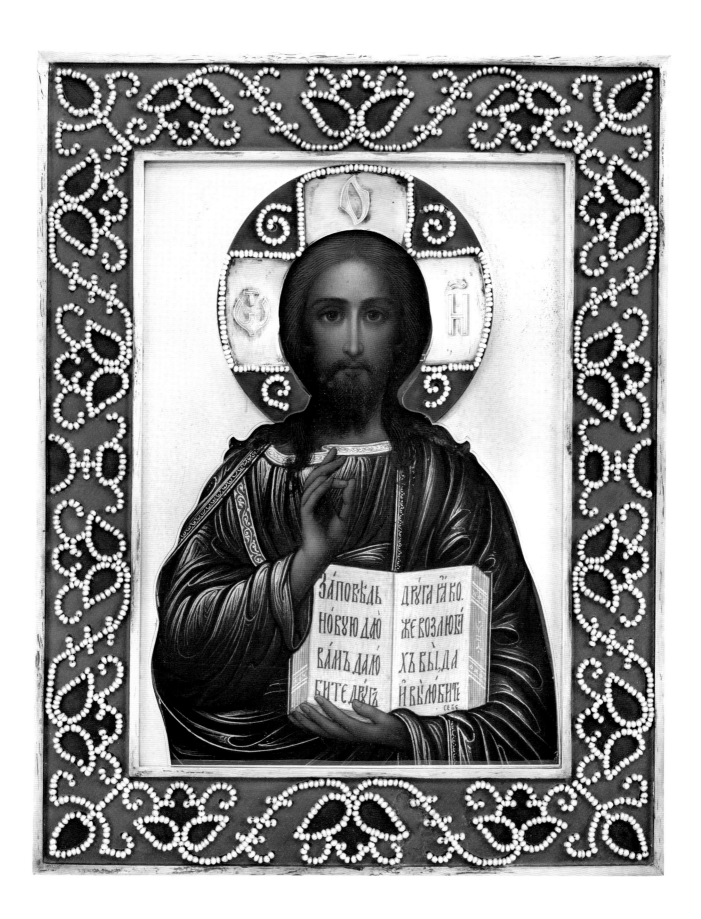

226

《容光焕发的基督》

1899～1908年
法贝热公司
工匠大师：亚尔马·阿姆菲尔特
锡板油画、银、珐琅、祖母绿、蓝宝石、石榴石、木
高 241.3 毫米　宽 171.45 毫米　厚 15.875 毫米
弗吉尼亚美术馆藏
莉莲·托马斯·普拉特遗赠

226

Christ Transfigured

1899–1908
Fabergé firm
Workmaster: Hjalmar Armfelt
Oil on tin, silver, enamel, emeralds, sapphires, garnet, wood
9 1/2" H × 6 3/4" W × 5/8" D
Virginia Museum of Fine Arts
Bequest of Lillian Thomas Pratt

227

《奇迹创造者圣尼古拉》

19～20 世纪
俄罗斯
木板油画、金箔
高 314.325 毫米　宽 266.7 毫米　厚 38.1 毫米
弗吉尼亚美术馆藏
莉莲·托马斯·普拉特遗赠

227

St. Nicholas the Miracle Worker

19th–20th Century
Russian
Oil and gold leaf on wood
12 3/8" H × 10 1/2" W × 1 1/2" D
Virginia Museum of Fine Arts
Bequest of Lillian Thomas Pratt

| 228 | 228 |

《奇迹创造者圣尼古拉》

19 ~ 20 世纪
俄罗斯
木板油画、金箔、黄铜、天鹅绒
高 225.425 毫米　宽 180.975 毫米　厚 25.4 毫米
弗吉尼亚美术馆藏
莉莲·托马斯·普拉特遗赠

St. Nicholas the Miracle Worker

19th – 20th Century
Russian
Oil and gold leaf on wood, brass, velvet
8 7/8" H × 7 1/8" W × 1" D
Virginia Museum of Fine Arts
Bequest of Lillian Thomas Pratt

229	229
《全能的基督》	Christ Pantocrator
1914～1917年	1914–1917
法贝热公司	Fabergé firm
木板油画、银镀金、银、石榴石、蓝宝石、可能是黄玉、锆石、金绿宝石、钻石、珍珠、天鹅绒	Oil on panel, silver gilt, silver, garnets, sapphires, possibly topaz, zircon, chrysoberyl, diamonds, pearls, velvet
高 298.45 毫米　宽 260.35 毫米　厚 38.1 毫米	11 3/4" H × 10 1/4" W × 1 1/2" D
弗吉尼亚美术馆藏	Virginia Museum of Fine Arts
莉莲·托马斯·普拉特遗赠	Bequest of Lillian Thomas Pratt

230

《艾弗斯卡亚圣母》

1914 ~ 1917 年

法贝热公司

木板油画、银镀金、银、石榴石、蓝宝石、黄玉（?）、碧玺（?）、锆石、珍珠、天鹅绒

高 301.625 毫米　宽 260.35 毫米　厚 38.1 毫米

弗吉尼亚美术馆藏

莉莲·托马斯·普拉特遗赠

230

The Iverskaya Mother of God

1914–1917

Fabergé firm

Oil on panel, silver gilt, silver, garnets, sapphires, topaz（?）, tourmaline（?）, zircon, pearls, velvet

11 7/8" H × 10 1/4" W × 1 1/2" D

Virginia Museum of Fine Arts

Bequest of Lillian Thomas Pratt

231

《圣容显现、圣伊丽莎白、拉多聂兹的圣谢尔盖三联画》

1884 年
俄罗斯
木板油画、银、银镀金、珐琅
打开：高 184.15 毫米　宽 269.875 毫米　厚 11.113 毫米
弗吉尼亚美术馆藏
莉莲·托马斯·普拉特遗赠

231

The Transfiguration, St. Elizabeth, and St. Sergius of Radonezh Triptych

1884
Russian
Oil on wood, silver, silver gilt, enamel
Open: 7 1/4" H × 10 5/8" W × 7/16" D
Virginia Museum of Fine Arts
Bequest of Lillian Thomas Pratt

232

《喀山圣母、圣亚力山大·涅夫斯基大公、抹大拉的圣玛利亚三联画》

1891 年
法贝热公司
工匠大师：帕维尔·奥夫奇尼科夫
木板油画、银镀金、珐琅、珍珠、红宝石、祖母绿、钻石、蓝宝石
打开：高 311.15 毫米　宽 384.175 毫米　厚 22.225 毫米
弗吉尼亚美术馆藏
莉莲·托马斯·普拉特遗赠

232

The Holy Virgin of Kazan, St. Prince Aleksandr Nevskii, St. Mary Magdalene Triptych

1891
Fabergé firm
Workmaster: Pavel Ovchinnikov
Oil on panel, silver gilt, enamel, pearls, rubies, emeralds, diamonds, sapphires
Open: 12 1/4" H × 15 1/8" W × 7/8" D
Virginia Museum of Fine Arts
Bequest of Lillian Thomas Pratt

233

《耶稣复活三联画》

1899～1908 年
法贝热公司
金板油画、黄金、祖母绿、蓝宝石、红宝石、钻石、珍珠
高 103.188 毫米　宽 150.813 毫米　厚 12.7 毫米
弗吉尼亚美术馆藏
莉莲·托马斯·普拉特遗赠

233

The Resurrection Triptych

1899–1908
Fabergé firm
Oil on gold, gold, emeralds, sapphires, rubies, diamonds, pearls
4 1/16" H × 5 15/16" W × 1/2" D
Virginia Museum of Fine Arts
Bequest of Lillian Thomas Pratt

234

圣乔治屠龙吊坠

1908～1917年
法贝热公司
银镀金、珐琅、红宝石、雀眼枫木
高 114.3 毫米　宽 85.725 毫米　厚 9.525 毫米
弗吉尼亚美术馆藏
莉莲·托马斯·普拉特遗赠

234

Pendant of St. George Slaying the Dragon

1908–1917
Fabergé firm
Silver gilt, enamel, rubies, birdseye maple
4 1/2" H × 3 3/8" W × 3/8" D
Virginia Museum of Fine Arts
Bequest of Lillian Thomas Pratt

专 论

Essays

伪法贝热

格扎·冯·哈布斯堡

伪法贝热（Fauxbergé）是笔者杜撰的一个名字，用于指代自20世纪30年代起涌入西方艺术市场的不计其数的法贝热（Fabergé）模仿品或人造（伪造）品[1]。其中，有很多都流入了收藏有大量法贝热作品的美国公共收藏机构。莉莲·托马斯·普拉特于1947年遗赠给弗吉尼亚美术馆的珍贵俄罗斯艺术藏品就是其中之一，如今，它们在美国同类藏品中占据着最重要的地位。这些艺术品诞生于1933年至1945年间；当时，凡能与俄罗斯帝国浪漫史扯上关系的物件在市场上必是抢手货——这也反映了供应商的狡诈：几乎每件出售的物品都附有一份书面声明，证明其"出身显赫"，曾经属于沙皇、其5个孩子或皇太后所有，或者至少来自"皇室成员"。从框画形式中，也能看出这种对来源的凭空捏造。例如，普拉特藏品的54幅框画中只有两幅为原始图片，其余装裱作品均为随机选取、任意剪贴的其他照片或皇室成员的复制品；如无可用照片，则使用报纸图片。所有这些都是精心伪造，只不过伪造的不是物品本身，而是所谓的来源。

包括卡地亚在内的一些公司还故意错误认定这些作品的出处，这也是司空见惯的做法。收藏家唯独钟情于法贝热的作品，以及曾属于惨遭杀害的皇室成员或其近亲的物品。这些收藏家中热情最高的当属4名美国女士——每获得一件物品，她们就感觉与罗曼诺夫家族更近了一步[2]。

卡尔·法贝热还在世时，对其作品的伪造就已经出现了。无论在国内还是国外，他的艺术作品都拥有难以计数的模仿者，而在当时没有人会仔细核查标志。1900年巴黎世博会期间，他展出的花器在德国俘获了众多拥趸[3]。法贝热本人也抄袭过18世纪的艺术作品，并在不法经营者的鼓动下签上自己的名字，以期以假乱真[4]。法贝热去世后不久，其子在巴黎继续生产俄罗斯风格的作品，有时竟会被误认为是原作。

俄国革命后，各地对法贝热作品的需求几乎都降为零；唯独英国除外，在那里法贝热仍是一个家喻户晓的名字，这是因为1903年至1915年间伦敦设有该公司唯一一家国外分支机构（公司办事处），而且英国王室成为了收藏主体。早在 1927 年，沃尔塔斯基公司的伊曼纽尔·斯诺曼就开始迎合这种客户需求，开拓性地展开俄罗斯采购之旅，将法贝热的创意艺术品及其他珍宝收入囊中。20 世纪 20 年代，法贝热的作品仍是公认的二手货；但到大萧条结束时（20 世纪 30 年代中期和末期），英国人和美国人已开始认真收藏。藏品供应十分充足，而年轻的苏联也很高兴甩掉这些令人讨厌的沙皇标志物。二战爆发后，法贝热真品从俄罗斯向西方的流入戛然而止，大量伪造品由此登场。在巴黎，两位天才伪造者——西尔维奥·马鲁切利和一位被称作什蒂科尔的先生——据称一直非常活跃，生产的艺术品质量极高。

1953 年，肯尼斯·斯诺曼出版了《卡尔·法贝热艺术品》（1962 年和 1964 年又先后再版），自此，有关法贝热的早期出版物便大量涌现，这进一步点燃了众多俄罗斯动物和人物（半）宝石雕刻伪造者的热情。莫斯科著名的伪造者大师、才华横溢的玉雕师瑙姆·尼古拉耶夫斯基和他在列宁格勒的内弟瓦西利·科诺瓦连科[5]向驻苏的西方和非洲外交官兜售了大量真品的仿制品，这些人厚颜无耻地利用外交渠道将所获之物走私到了西方。伪造者还利用不知内情的拍卖行——尤其是美国的拍卖行——来出售他们的产品。

此外，在列宁格勒，成功人士米哈伊尔·莫纳斯特尔斯基[6]成了 20 世纪 70 年代中期至 90 年代玉雕伪造品的主要源头。他的手艺甚至骗过了俄罗斯的各大博物馆。他的雕刻要么按照斯诺曼的著作，直接模仿独一无二和众所周知的艺术品，要么是自己发明一些充满"民间气息"的人物角色或童话动物。其中比较典型的是一位俄罗斯流亡者，他是莫纳斯特尔斯基的众多受骗者之一，花费了大量金钱打通关节，将成箱的动物玉雕和人物雕像走私出了苏联。20 世纪 80 年代，他的"藏品"常在瑞士成批出售，这恰好欺骗了阿曼德·哈默——他希望为查尔斯王子的联合世界书院谈成这笔交易。

法贝热的伦敦代表亨利·班布里奇——他一定经手过现在为伊丽莎白女王所有的 25 件花器中的大部

分——就曾警告过，要辨别这类艺术品的伪造品难度非常大。1949年，班布里奇就美国的伪造品写道：

> （法贝热）的花器如此成功，收藏家必须擦亮眼睛，因为其仿造品难以计数，购买时一定要格外小心……法贝热花器的收藏家切不可仅依靠商标和制造商的标志来判别，如果他们对其主要特点缺乏足够了解——内行人一眼就能看出这些特点营造出的整体感觉——那么就应该寻求指导……真品实际上供不应求。对法贝热作品尚无鉴赏能力的收藏家，应向信誉好的经销商寻求指导[7]。

这是班布里奇多年前的告诫，今天仍完全适用：

> 务必牢记原作花器的精湛抛光，在这方面要始终保持一种批判意识。要仔细检查花柄，并注意大多数情况下它们的线纹分布得非常均匀，即使线纹有时几乎看不出来。要注意花柄表面呈现不同程度的粗糙状。要注意检查叶子的背面和正面，其叶脉应为上下对应。对于美得难以形容的花朵，唯一要做的就是拿一个显微镜，跪下来仔细检查它的每处细节。天然花朵的那种精致与复杂获得了忠实再现；受到所采用材料和必要机械装置的限制，这种逼真达到了不可复制的程度。

法贝热的花器非常罕见，专家对此似乎已有共识：有些人声称至少存在60朵，还有人说是80朵[8]。因此，1938年至1942年间，美国市场上出现了60朵以上，这至少可以说是值得怀疑的。皇家收藏中的25朵花器是现存最大一批同类藏品[9]，不应以此来衡量和判断其他花器[10]。按照班布里奇在1949年的告诫，20世纪40年代及以后进入皇家收藏的虽然几乎可以肯定是真品，但至少也应进行仔细查验。

法贝热真品切花总是充满了自然气息，它们一般为单株，以一定的角度倾斜；花柄"连接"在水晶花瓶的底部并斜靠在花瓶边缘——从来都不是直立的，花茎也没有"半浮"在假水中。绝大多数的早期花器都拥有盛开的花朵、金色的花茎和软玉做的叶子，在水晶花瓶中挺身而立[11]。由于叶柄的脆弱性，它们一般都没有标记[12]。在那些载有标志的作品中，只有两朵带有工匠大师亨里克·维格斯顿的首字母，其制造年份可推测为1903年至1917年之间。鲍文玉（硬绿蛇纹石）或软玉花瓶提供了几个实例——且大多是后期的花器。一本附有法贝热原作照片的相册展示了日本风格的后期花器，许多都装在带有仿真金土的花盆内，或是摆放在日本风格的基座上。

美国的法贝热花器大都来自阿曼德·哈默[13]（1898～1990年），他是一位俄罗斯犹太移民的儿子。1921年，哈默第一次前往苏联，当时美国与俄罗斯还不能自由通商。他成为了美国农业机械的采购代理，还是30多家美国公司在苏联的独家代理。1925年至1930年间，他掌握着一家铅笔制造业的特许权，据称总收入可达850万卢布，约合400万美元。工厂被没收后，阿曼德和他的弟弟——古董商维克多·哈默显然获得了苏联官员的授权，独家代表他们向美国出售俄罗斯帝国的宝藏。兄弟二人在经济萧条时期抵达纽约，带来了满箱的俄罗斯艺术品和纪念品——其中包括10只帝国复活节彩蛋，开始在美国各地进行推销。1932年，他们在23个城市的大型百货公司组织了为期6个月的巡展。1933年至1934年，哈默兄弟在纽约市东52街安顿下来，开始在罗德与泰勒百货店进行"俄罗斯皇家展"半永久式展览，合同期3年。芝加哥世博会举行期间，他们还在芝加哥的马歇尔广场商店展出他们的珍宝。自此，他们的事业不断发展壮大。负责苏联对外贸易的阿纳斯塔斯·米高扬源源不断地向兄弟二人提供俄罗斯的文物，直到1939年9月二战爆发，他们的主要货源才枯竭。

哈默兄弟与A La Vieille Russie的亚历山大和雷·谢弗共同打开了美国的法贝热市场（特别是花器），此后便开始寻找新的货源和获利手段。新一代花器的来源尚不清楚。它们也许产自德国的伊达尔-奥伯施

图1：法贝热，野玫瑰胸针，约1910年；黄金、珐琅；直径1 1/2英寸（2.8厘米）。莉莲·托马斯·普拉特遗赠，弗吉尼亚美术馆藏。

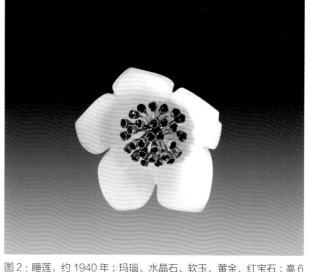

图2：睡莲，约1940年；玛瑙、水晶石、软玉、黄金、红宝石；高6 3/8英寸（16.2厘米）。莉莲·托马斯·普拉特遗赠，弗吉尼亚美术馆藏。

泰因，那里是西欧的主要玉雕中心。维克多·哈默提到的另一个渠道是巴黎，据说那里的工匠能够使用苏联提供的一套法贝热原始制作工具"打造出新品"。班布里奇在1949年写道："这种恶意模仿带有预谋性质，每朵花的金柄上都加盖有伪造的法贝热标记和俄罗斯的黄金标志。"[14]

在弗吉尼亚美术馆普拉特藏品的22朵花器中，其中只有一件——即购买于1935年的真品矢车菊——是由普拉特夫人在1939年1月9日至1940年1月9日之间从哈默手中购得的。其中有15件都可能是伪造品。它们的风格各不相同：有的是精雕细琢，也许出自法国工艺；而有些则较为粗糙，可能源于德国。但是，它们都有法贝热以及后来添加的魏格斯特姆标志。

美国的其他博物馆和私人藏家也收藏了与普拉特藏品相似的花器。哈默的经商魅力和营销技巧吸引了一批富有的收藏家。其中一位便是哈里森·威廉姆斯酷爱时尚的妻子——莫娜·威廉姆斯，即后来的莫娜·冯·俾斯麦伯爵夫人。她被法国著名女装设计师评为"全球最佳着装女性"，其巴黎沙龙的一个柜子内装满了伪造花器。所有花器都带有标记，装花的小盒子酷似法贝热作品，并带有哈默的地址（据笔者回忆，在她布置优雅的沙龙中，共收藏了大约20件这类花器）。

请牢记亨利·班布里奇对真品花器的详细介绍，大部分哈默的花器看上去都了无生气；其质量较差，雕刻简单，雌蕊花丝过于死板（对比图1和图2）。普拉特藏品中的花器除6件外，其他都以西里尔字母标有法贝热、工匠大师亨里克·维格斯顿以及12索拉尼金等字样。每朵花上也有同样的标志（例外包括图版129的铃兰，其上有伪造的法贝热阴刻签名；图版132也是伪造的三色堇，它可能有一个标志被叶片遮住了）。其余的花都被认为是真品，金凤花（图版130）带有法贝热的莫斯科标志；而矢车菊、紫罗兰（图版137）和三色堇都没有标志。

普拉特藏品总共包括34件动物玉雕、1个银兔水壶（正品）和3只由银或金制作的大象（伪造）。

然而，新证据不禁让人怀疑这种归属的正确性；有1300件类似于卡德尔和至少5位圣彼得堡同时代雕塑家的作品，这让问题变得更加复杂。15件用半宝石制作的动物雕塑在此被认为属于法贝热；9件属于卡德尔（5件动物雕塑载有伪造的法贝热标志，在此将其归属于卡地亚）；4件系由其他俄罗斯工匠出品。

莉莲·普拉特的54件宝贝要好辨认得多。其中，有42件明显是正品，2件可归属于法贝热，5件是其他大师的作品或没有标记，只有两件是伪造的。但不要忘记，就图片内容及其来源的不真实性而言，所有54件作品的信息实际上都存在误传。有两件伪造品的质量非常差，并草草刻有法贝热竞争对手卡尔·哈恩的标志。它们缺少卡尔·布兰克或亚历山大·特莱登这两位工匠大师的必备标志，其首字母出现在哈恩的所有后期作品上。该相框饰有一个宝石双头鹰俄罗斯国徽，以表明其帝国身份。然而，这种国徽只会出现在烟盒和小珠宝上：帝国内阁的货单中从未出现过带照片的相框[15]。其他现存的相框——两个由法贝热制作，其他出自博林和克希利之手——拥有手绘的沙皇微缩模型。这些虚构的照片只能进一步证实一个事实，即这两件都是革命后的伪造品，可能出自法国人之手。

阿曼德·哈默向莉莲·普拉特称赞了卡尔·哈恩公司，说其与法贝热一样声名显赫。有趣的是，俄亥俄州克利夫兰美术馆和巴尔的摩沃尔特斯艺术博物馆的两个帝国鼻烟壶据称是法国制造，其中前者来自哈默，但也带有可疑的哈恩标志[16]。罗曼诺夫王朝300周年纪念相框也非常神秘，以前曾属于福布斯藏品[17]。这件豪华作品曾于20世纪30年代在美国市场出现，虽镶有同样的哈恩标志，日期却是1913年——要知道哈恩工作室在两年前就已关门。这些物品装饰着沙皇微缩模型和年轻的阿列克谢，但在遗留下来的帝国委员会文件中并没有记载。

普拉特藏品中最炫目的一件是一个镶嵌着宝石、涂着珐琅的圆形展盒（图3），它被作为法贝热的真品出售，其实是一件有趣的伪造品。其制造者并不熟

图3：宝石珐琅金盒；黄金、珐琅、钻石、红宝石、珍珠；直径 3 5/8 英寸（9.2 厘米）。莉莲·托马斯·普拉特遗赠，弗吉尼亚美术馆藏。

悉法贝热工作室采用的标记方法。这个盒子标有工匠大师米哈伊尔·普尔金（他只在圣彼得堡进行创作）的首字母，以及法贝热以西里尔字母出现的帝国授权标志"K. Fabergé"和俄罗斯双头鹰。但是，这个帝国标志只用于在莫斯科生产的作品。其珐琅内饰非常精细，两侧和底部均缺少常用的珐琅涂层，因此从设计风格判断，它更接近法贝热竞争对手的作品。因此，这也是一个错误认定。

注释：

1. 有关法贝热伪造品的更多资料，见"哈布斯堡和索洛德科夫"，1979 年，页 149-150；"哈布斯堡"，1986 年，页 334-335；"哈布斯堡和洛帕拖"，1993 年，页 165-166。"哈布斯堡"，1996 年，页 329-338；"索洛德科夫"，1995 年，页 42-43；"哈布斯堡"，2000 年，页 385-388。

2. 美国的四大收藏家分别是新奥尔良的玛蒂尔达·戈丁斯·格雷；华盛顿伍德希尔的玛荷丽·梅莉薇德·波斯特；俄亥俄州克利夫兰的因迪亚·厄尔利·明歇尔；弗吉尼亚州的莉莲·托马斯·普拉特。

3. 有关德国和奥地利的伪造花器，可参见"哈布斯堡和洛帕拖"，1993 年，页 458。

4. 同上，446。

5. "布罗克琳"，1975 年，页 163-185。

6. "哈布斯堡"，2000 年，页 386-387。

7. "班布里奇"，1968 年，页 105-106。

8. "斯威齐"，2004 年。伦敦销售分类账列出了 1 万多件售出物品中的 40 件花器。俄罗斯的相关研究表明，1896 年至 1908 年只有 6 件、1908 至 1915 年只有 8 件花器被女皇购得。

9. "吉托"，2003 年，页 102-121。

10. 在英国女王伊丽莎白二世所拥有的花器中，有 5 件可追溯至法贝热的伦敦销售店；7 件由亚历山德拉王后购得，日期不详；11 件"可能由亚历山德拉王后购得"，首次在 1953 年的藏品中被提及；1 件在 1934 年由玛丽女王购得；还有两件分别购于 1944 年和 1947 年。

11. 英国皇家收藏的 25 件花器中，有 22 件是立在水晶花瓶中的，而且除 1 件外，其他全部是珐琅花瓣。据班布里奇所言，这些都是公司首席珐琅工匠亚历山大·尼古拉·彼得罗夫和博伊佐夫的作品。在玛利亚·帕夫洛夫娜大公夫人所拥有的 34 件花器中，有 23 件是立在水晶花瓶内的。2004 年斯威齐中所画的 62 件花器中，有 56 件是立在水晶花瓶内的。另一方面，在美国公共藏品中的 40 多件花器中，有 32 件是玉雕花瓣，三分之二立在玉盆内。伯恩鲍姆描述道："蒲公英制作得尤其成功：它们的绒毛非常自然，金线上镶嵌着小巧玲珑、未经切割的钻石。这些钻石在白色绒毛中不断闪烁，这是它最成功的一点，从而避免使这种假花显得过于矫揉造作。"

12. 在伊丽莎白二世女王收藏的 25 件花器中，只有 8 件带有标记，除两件外均出自工匠大师亨里克·维格斯顿之手（皇家收藏中的一个例外是柔荑花，其上刻有法贝热金匠费多尔·阿法纳斯的首字母，这也是流传下来的唯一一件）。2004 年斯威齐中所画的 62 件花器中，只有 12 件带有标记，其中的几件在此被认为是伪造品。

13. "爱泼斯坦和哈默"，1996 年，页 135-142。有关阿曼德·哈默对法贝热的收藏，可参见"布鲁梅和爱德华兹"，1992 年，页 95-108；"哈布斯堡"，1996 年，页 57-61。

14. "班布里奇"，1968 年，页 105。

15. "提兰德–戈登西乐姆"，2005 年，页 149-178 和页 179-190。

16. 克利夫兰美术馆，因迪亚·厄尔利·明歇尔遗赠（"霍利"，1967 年，91，编号 43），一个软玉盒以钻石镶边，装饰有沙皇尼古拉二世和亚历山德拉·费奥多罗芙娜皇后的微型画像，带有 1838 年的巴黎限制授权标志以及 1899 年之前的圣彼得堡标志；巴尔的摩沃尔特斯艺术博物馆，亨利·C·沃尔特斯的遗赠。莉莲·托马斯·普拉特藏品中带有俄罗斯国徽的粉红色珐琅相框也拥有相同的小担保标志。

17. 该相框出现在"福布斯和特罗默"中，1999 年，页 261。

Fauxbergé

Geza Von Habsburg

Fauxbergé is a word coined by this author to denote the myriad of Fabergé lookalikes, or faux (forged) objects, that have inundated the Western art market since the 1930s.[1] A number of these have landed in American public collections with substantial Fabergé holdings. The great Russian art collection bequeathed by Lillian Thomas Pratt to the Virginia Museum of Fine Arts in 1947 — today the most important of its kind in the United States — is no exception. Formed between 1933 and 1945, when the romance of a Russian imperial provenance was deemed an essential prerequisite for a sale, it reflects the cunning of the suppliers: virtually each object sold came with a written statement declaring an august provenance of having originally belonged to either the tsar or tsaritsa, one of their five children, or the dowager empress — or at least a "Member of the Imperial Family." Further evidence of such manipulated origins is available in the form of framed images. For example, in the Pratt collection only two of fifty-four frames contain their original photographs. The remainder hold other photographs or reproductions of imperial family members, chosen at random and cut to size, or, if such photographs were not readily available, newspaper images. All this, too, was a form of sophisticated forgery, not of the objects themselves but of their purported origins. The willful misattribution of objects by other companies, including Cartier, was a common practice. Passionate collectors demanded objects by Fabergé only and those that had belonged to a member of the tragically murdered imperial family, or to one of their closest relatives. It would seem that the most ardent of these collectors were four American women who felt more personally linked to the Romanovs with each and every object they acquired.[2]

The forging of Fabergé objects began during Karl Fabergé's own lifetime. His works of art found numerous close imitators, at home and abroad, at a time when hallmarks were not yet scrutinized. As early as the 1900 Universal Exposition in Paris, the flowers he exhibited apparently engendered quite a following in Germany.[3] When Fabergé himself convincingly plagiarized eighteenth-century works of art, he, too, was encouraged by unscrupulous operators to leave off his signature in order to allow his version to be passed off as the genuine article.[4] Shortly after his death, his sons in Paris continued the production of objects in the Russian style, which have sometimes been mistaken for originals.

There was virtually no demand for Fabergé immediately after the Russian Revolution, except in Britain. Fabergé was still a familiar word there thanks to the existence of a London office between 1903 and 1915 (the firm's only branch abroad) and the British royal family's leading role as collectors. Emmanuel Snowman of the Wartski firm catered to this clientele as early as 1927, acquiring seminal works of art and other treasures by Fabergé during his pioneering buying trips to Russia. If Fabergé objects in the 1920s were still generally considered as secondhand goods, by the end of the Great Depression, in the mid-to-late 1930s, both the British and Americans began collecting seriously. Supply was plentiful, and the young Soviet Union was glad to be rid of all the trappings of the hated tsarist regime. World War II put a stop to the influx of genuine Fabergé objects from Russia to the West and marked the beginnings of serious forgery. In Paris two talented forgers, Silvio Marucelli and a certain Monsieur Stiquel, are alleged to have been active, producing objects of high quality.

The spate of early publications on Fabergé, beginning with Kenneth Snowman's 1953 *The Art of Carl Fabergé* (and its enlarged editions of 1962 and 1964), were instrumental in inspiring a number of Russian forgers of hardstone animals and figures. The gifted stone carver Naum Nikolaevskii, a celebrated master forger in Moscow, and his brother-in-law Vasilii Konovalenko in Leningrad[5] sold copies of genuine one-of-a-kind objects on a large scale to Western and African diplomats in the Union of Soviet Socialist Republics who shamelessly used the diplomatic pouch to smuggle their acquisitions to the West. Forgers also made use of unsuspecting auction houses, particularly those in the United States, to offload their wares.

Also in Leningrad, the highly successful Mikhail Monastyrskii[6] was the source of a great number of hardstone forgeries from the mid-1970s to the 1990s. Even major

Russian museums were taken in by his craftsmanship. His carvings were either straight copies of one-of-a-kind, well-known subjects derived from Snowman's books or his own inventions of "folksy" characters or fairy-tale animals. Typical is the case of a Russian émigré, one of Monastyrskii's numerous dupes, who had paid his way out of the Soviet Union and smuggled suitcases full of hardstone animals and figurines. His "collection" was frequently offered as a unit in Switzerland in the 1980s, deceiving none other than Armand Hammer, who hoped to negotiate its sale in order to benefit Prince Charles's United World Colleges.

Henry Bainbridge, Fabergé's London representative, who therefore must have actually handled many of the twenty-five flowers now owned by Queen Elizabeth, warned of the complexities inherent in identifying any forgeries in that genre. In 1949, with the American forgeries in mind, Bainbridge wrote:

So successful was [Fabergé] with his flowers, that it behooves collectors to see that they get the right thing, for they have been imitated, not in great numbers, but to such an extent as to call for care when purchasing them. . . The collector of Fabergé flowers must forget hall-marks and maker's marks altogether, and if he is not sufficiently acquainted with their chief characteristics, which in toto make up something which to the experienced eye can be seen at a glance, then let him seek guidance. . .genuine pieces are in short supply. Collectors who have not developed a Fabergé sense of their own should be guided by the advice of dealers of repute.[7]

Bainbridge's admonitions of years ago remain fully valid today:

Always keep in mind the superb finish of the original natural flower and keep the critical sense at this pitch. Examine the stalks minutely and note that for the most part they . . . are evenly lined, even if sometimes the lines are hardly to be seen. Let him notice that the surface of the stalks is, in varying degree, mat. Let him examine the back of the leaves, as well as the front and take note that the veinings are correspondingly ribbed on the back. Of the flower blooms one can only say that their delicacy is indescribable, the only thing he can do is to take a microscope and on bended knee saturate himself with their unending detail. . .[T]he delicacy and intricacy of the natural flower has been faithfully reproduced up to the point where further reproduction is no longer possible owing to the limitations of the material employed and the mechanical devices necessary.

Specialists seem to agree that Fabergé flowers were rare; some claim as few as sixty were made while others say eighty.[8] Thus, the appearance of over sixty flowers on the American market between 1938 and 1942 is, to say the least, somewhat suspect. The Royal Collection, with its twenty-five flowers, is by far the largest extant group of its kind[9] and should be used as a benchmark by which to judge other flowers.[10] Because of Bainbridge's warnings in 1949, even the flowers that entered the Royal Collection in the 1940s and later, though almost certainly genuine, should at least be scrutinized.

Genuine Fabergé cut flowers are always depicted naturalistically, generally as single sprays leaning at an angle, the stalk "grounded" in the rock-crystal vase's base and resting against the rim of the vase—never standing upright and never with stems "floating" halfway in the simulated water. The large majority of early flowers have enameled blossoms, gold stems, and nephrite leaves and stand in transparent rock-crystal vases.[11] They are generally unmarked, because of the fragility of the stalks.[12] Of those that are hallmarked, all but two bear the initials of head work master Henrik Wigström, dating them to 1903-17. A few, mostly later, examples are set in bowenite or nephrite vases. An album with original Fabergé photographs shows late flowers in the Japanese taste, many in hardstone pots with simulated gold soil or on Japanese style bases.

The primary source of Fabergé flowers in the United States was Armand Hammer(1898-1990),[13] son of a Russian Jewish emigrant. Hammer first traveled to the USSR in

1921, when American contacts with Russia were still forbidden. He became the purchasing agent for American agricultural machinery and exclusively represented up to three dozen U.S. companies in the Soviet Union. Between 1925 and 1930 he owned a pencil-manufacturing concession, which is said to have grossed 8.5 million rubles, about $4 million. Dispossessed of his factory, Armand and his brother, antique dealer Victor Hammer, were apparently authorized by the Soviet officials to represent them as sole agents for the disposal of Russian imperial treasure in the United States. Arriving in New York in the middle of the Depression, laden with trunks full of Russian artworks and memorabilia including ten imperial Easter eggs, the entrepreneurial brothers set about marketing their wares across America. They began by improvising a six-month traveling show in 1932, which they staged at the leading department stores in twenty-three cities. The Hammers settled in New York City at 3 East Fifty-second Street with a three-year contract for a semipermanent display entitled *Russian imperial Exhibit* at the city's Lord & Taylor store in 1933-34. They also exhibited their treasures at Marshall Fields in Chicago, concurrent with the Chicago World's Fair. Business thrived. The USSR's commissar for foreign trade, Anastas Mikoian, saw to it that the brothers were kept supplied with Russian artifacts, until World War II broke out in September 1939 and their primary source dried up.

Having created, together with Alexander and Ray Schaffer of A La Vieille Russie, a demand for Fabergé (in particular the floral compositions) in the United States, the Hammers embarked upon finding new sources and means of continuing their lucrative business. The origin of this new generation of flowers is yet unclear. Perhaps they were produced in Idar-Oberstein in Germany, Western Europe's chief hard-stone-carving center. Another option, mentioned by Victor Hammer, is Paris, where a craftsman was said to have been capable of "knocking off new objects" that were then hallmarked with a set of original Fabergé tools supplied by the Soviets. In 1949 Bainbridge wrote: "the imitating was done with malice aforethought, the flowers in every case being stamped on the gold stalk with the forged mark of

Fig.1 : Fabergé, Wild Rose Brooch, ca. 1910; gold, enamel; diameter 1 1/2 in. (2.8 cm). Bequest of Lillian Thomas Pratt, Virginia Museum of Fine Arts.

Fig.2 : Water Lily, ca. 1940; agate, rock crystal, nephrite, gold, rubies; height 6 3/8 in. (16.2 cm). Bequest of Lillian Thomas Pratt, Virginia Museum of Fine Arts.

Fabergé and the Russian hallmark for gold. "[14]

Of the twenty-two flowers in VMFA's Pratt collection, all but one-the genuine *Cornflower Spray* purchased in 1935—were acquired by Mrs. Pratt between January 9, 1939, and January 9, 1940, from Armand Hammer. Fifteen of these are most probably forgeries. Their styles vary: some are more finely crafted, perhaps denoting a French origin, while others are cruder, possible indicating a German provenance. However, they all share the Fabergé and Wigström hallmarks that were added later.

Flowers similar to those in the Pratt collection exist in other museums and private holdings in the United States. Hammer's charm and clever marketing attracted a number of well-heeled collectors. One of these was the fashionable wife of Harrison Williams, Mona Williams, later Countess von Bismarck. Named the "Best Dressed Woman in the World" by the leading French couturiers, she had a cabinet full of such forged flowers in her Paris Salon. All were hallmarked, and their fitted cases were Fabergé look-alikes stamped with Hammer's address (this author recalls some twenty examples in her elegant salon).

Keeping in mind Henry Bainbridge's enthusiastic remarks about genuine flowers, the majority of the Hammer flowers look lifeless; their quality is poor, their carving summary, and their wire pistils rigid (compare figs. 1 and 2). With six exceptions, all of the Pratt flowers bear identical marks of Fabergé in Cyrillic, that of head work-master Henrik Wigström, as well as the *72-zolotnik* gold mark. The same hallmarking tools were used for each flower (the exceptions are the lilies of the valley, cat. no. 129, which bears a forged Fabergé incised signature; and a forged pansy, cat. no. 132, which may have a mark hidden by leaves). Of the remaining flowers, accepted here as genuine, *Globeflowers* (cat. no. 130) bears Fabergé's Moscow hallmark, and a cornflower; a violet (cat. no. 137); and a pansy are unmarked.

The Pratt Collection includes thirty-four hardstone animal figures together with a silver rabbit pitcher (genuine) and three elephants (forgeries) made of silver or gold. However, new evidence calls into question the validity of their attributions, further complicated by 1,300 known Cartier look-alikes and the close proximity to Fabergé of sculptures by at least five St. Petersburg contemporaries. Fifteen of the animal figures in semiprecious stones are considered here to be by, or attributed to, Fabergé; nine by Cartier (five of the animals bear forged Fabergé marks and are attributed here to Cartier); and four to other Russian craftsmen.

Lillian Pratt fared much better with her fifty-four frames. Of these, forty-two are clearly genuine, two can be attributed to Fabergé, five are by other masters, or unmarked, and only two are forgeries. It should not be forgotten, however, that virtually all fifty-four frames are misrepresented as far as their photographic contents and their spurious provenances are concerned. The two forged frames are of poor quality and bear the unevenly struck hallmarks of Fabergé's competitor Karl Hahn. They lack the obligatory marks of one of his two leading workmasters, Carl Blank or Alexander Treyden, whose initials appear on all of Hahn's later productions. The frames are applied with the Russian state emblem in the form of a jeweled double-headed eagle, allegedly denoting an imperial presentation. This emblem, however, only appears on cigarette cases and small jewels: no presentation frames with photographs are recorded among the invoices of the Imperial Cabinet.[15] The other extant presentation frames, two by Fabergé and the others by Bolin and by Koechly, have hand-painted miniatures of the tsar. The apocryphal photographs only further corroborate the fact that both objects are of postrevolutionary, probably French, fabrication.

Armand Hammer praised the firm of Karl Hahn to Lillian Pratt as equal in fame to Fabergé. Interestingly, two allegedly French-made imperial snuffboxes in the Cleveland Museum of Art in Ohio and the Walters Art Gallery in Baltimore, the former of which has a Hammer provenance, also bear questionable Hahn hallmarks.[16] Mystery also surrounds the so-called Romanov Tercentenary Frame, formerly in the Forbes collection.[17] This lavish object, which appeared on the American market in the 1930s, has the same Hahn hallmark but is dated 1913, two years after the Hahn workshop closed. No such object, set with miniature

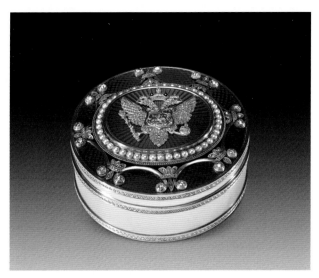

Fig. 3 : Jeweled and Enameled Gold Box; gold, enamel, diamonds, rubies, pearls: diameter 3 5/8 in. (9.2 cm). Bequest of Lillian Thomas Pratt, Virginia Museum of Fine Arts.

portraits of the tsar, tsaritsa, and young Aleksei, is among the surviving files of imperial commissions.

One of the showiest objects in the Pratt collection, a circular jeweled and enameled presentation box (fig.3) sold as a Fabergé original, is an interesting forgery. Its maker was unaware of the hallmarking system used in the Fabergé workshops. The box is struck with the initials of workmaster Mikhail Perkhin (who practiced exclusively in St. Petersburg) as well as Fabergé's imperial warrant mark. "K. Fabergé" in Cyrillic and the Russian double-headed eagle. This imperial mark was used only for objects produced in Moscow. With its finely enameled interior cover and its sides and base lacking their usual enamel coating, this box is stylistically closer to works by Fabergé's competitors. It is therefore at best a misattribution.

NOTES

1. Fabergé forgeries are discussed at length in Habsburg and Solodkoff 1979, 149-50; Habsburg 1986,334-35ff.; Habsburg and Lopato 1993, 161-66ff.; Habsburg 1996, 329-38; Solodkoff 1995,42-43ff.;Habsburg 2000,385-88.
2. The quartet of American collectors was Matilda Geddings Gray of New Orleans; Marjorie Merriweather Post of Hillwood,Washington, D.C.; India Early Minshall of Cleveland, Ohio; and Lillian Thomas Pratt of Virginia.
3. For German and Austrian forgeries of flowers, see Habsburg and Lopato 1993, 458.
4. Ibid.,446.
5. Brokhin 1975,163-85.
6. Habsburg 2000, 386-87ff.
7. Bainbridge 1968, 105-6.
8. Swezey 2004. The London sales ledgers list forty flowers out of a total of some ten thousand items sold. Research in Russia has shown that between 1896 and 1908 only six flowers were acquired by the empresses

and between 1908 and 1915 only eight more.

9. Guitaut 2003, 102-21.

10. Of Queen Elizabeth II's flowers five can be traced to Fabergé's London sales room; seven were bought by Queen Alexandra, date unknown; eleven were "probably acquired by Queen Alexandra," first mentioned in the collection in 1953; one was acquired by Queen Mary in1934; and two were acquired from Wartski in 1944 and 1947 respectively.

11. Of the twenty-five flowers in the British Royal Collection, twenty-two stand in rock-crystal vases, and, with one exception, all have enameled petals. According to Bainbridge these were the work of the firm's chief enamellers, Aleksandr and Nikolai Petrov and Boitsov. Of the thirty-four flowers owned by Grand Duchess Maria Pavlovna, twenty-three stood in rock-crystal vases. Of the sixty-two flowers illustrated in Swezey 2004, fifty-six stand in rock-crystal vases. On the other hand, of the over forty flowers in American public collections, thirty-two have hardstone blossoms and two-thirds stand in hardstone pots. Birbaum recorded that "the dandelions were particularly successful: their fluff was natural and fixed on with a golden thread with a small, uncut diamond. The shining points of the diamonds among the white fluff were marvelously successful and prevented this artificial flower from being too close to a reproduction of nature."

12. Of the twenty-five flowers in the collection of Queen Elizabeth II, only eight are hallmarked, with two exceptions, all by head workmaster Henrik Wigström (one exception in the Royal Collection is the *Catkin*, which as the sole apparent surviving example, is signed with the initials of Fabergé goldsmith Fedor Afanas'ev). Of the sixty-two flowers illustrated in Swezey 2004, only twelve are hallmarked, of which several are here considered to be forgeries.

13. Epstein and Hammer 1996, 135-142. For Armand Hammer's involvement with Fabergé, see Blumay and Edwards 1992, 95 108; Habsburg 1996, 57-61.

14. Bainbridge 1968, 105.

15. Tillander-Godenhielm 2005, 149-78 and 179-90.

16. Cleveland Museum of Art, bequest of India Early Minshall (Hawley 1967, 91, cat. 43), a nephrite box set with diamond-encircled, crowned miniature portraits of Tsar Nicholas II and Tsaritsa Alexandra Feodorovna, bearing a Paris restricted guarantee mark for 1838 through the present as well as marks of St. Petersburg before 1899; Walters Art Gallery Baltimore, bequest of Henry C. Walters. The pink enamel frame with the Russian emblem in the Lillian Thomas Pratt Collection also bears the same *petite garantie* mark.

17. The frame is illustrated in Forbes and Tromeur 1999, 261.

巧夺天工，至精至美——法贝热装饰艺术

宋海洋

2011年5月14日，故宫博物院与美国弗吉尼亚美术馆签署了两馆合作框架意向书，极大地促进了双方在主要藏品、人员交流和文化共享上的合作。2014年10月18日，筹备了3年多的"紫禁城——北京故宫博物院藏皇家珍品展"在弗吉尼亚美术馆隆重开幕，来自故宫博物院的119件（套）中国明清时期的皇室珍品向美国民众深入展示了绚丽璀璨的中国皇家艺术。2016年4月故宫博物院迎来了弗吉尼亚美术馆珍藏的234件（套）法贝热艺术珍品，举办"玲珑万象——来自美国的俄罗斯皇家法贝热装饰艺术展"，这将使中国观众在故宫博物院的展厅里欣赏到来自美国博物馆的精美藏品。

彼得·卡尔·法贝热（1846年5月30日～1920年），是19世纪末20世纪初俄罗斯著名的珠宝工艺美术设计大师。法贝热早年游历德、意、法、英等欧洲国家，1870年，20多岁的法贝热子承父业涉足制造珠宝工艺饰品。在1882年莫斯科举办的全俄展览会上，他的产品一举成名，成为欧洲各国皇家追捧的焦点。法贝热是一位杰出的工艺大师，他大胆革新，在莫斯科、基辅和伦敦开设的工作坊里，潜心设计雕琢了一批非同凡响的工艺美术品，其中尤以复活节彩蛋最负盛名，被称为"俄罗斯彩蛋"。彩蛋体现了俄国罗曼诺夫王朝的辉煌，至精至美，获得了俄国等各国皇室的广泛认同，被视为艺术珍品。

一　东正教与彩蛋艺术

东正教，又称正教、希腊正教，是基督教中的一个派别，主要依循东罗马帝国所流传下来的基督教传统。988年弗拉基米尔大公将东正教引入俄罗斯，自此之后东正教对俄罗斯政治、社会、文化、建筑、工艺等方方面面均产生了深远的影响。

复活节是俄罗斯东正教教历中最重要的节日，是纪念耶稣基督在十字架受难后第三天复活的节日，象征着春回大地、万物复苏。复活节的庆祝活动包括礼拜，传统的节日问候以及丰盛的宴请，当然还

有互赠神圣的复活节彩蛋，这是复活节中最具代表性的吉祥物。12世纪时，复活节中出现了用彩色涂画的彩蛋，寓意"耶稣复活，走出石墓"，获得重生与希望（李友友：《俄罗斯民间玩具文化初探》）。一枚复活节彩蛋，是基督耶稣复活以及人类赎罪的象征，是尘世生活与基督创造的天堂化身。甚至在基督诞生前的时代，人们也视鸡蛋为富饶的象征。彩蛋缤纷的颜色传达了阳光的闪耀和温暖，以及经历单调冬天后恢复生机自然的那种丰裕状态。俄罗斯人称复活节彩蛋为"pysanka"，绘制的颜料由洋葱、甜菜根、棉花及其他生活中常用的材料制成。除了简单的彩蛋，还有木质、玻璃、瓷、青铜彩蛋和珍贵石材与金属混制而成的彩蛋。

对于俄罗斯人来说，彩蛋能带来健康、美貌、力量和富足，是复活节庆典的必备之物。俄罗斯人对待彩蛋像对待圣像、保护神一样把它保存在家中，还以彩蛋来迎接新的生命。

二 法贝热——沙皇的御用珠宝商

彼得·卡尔·法贝热1846年出生在圣彼得堡，他的父亲古斯塔夫·法贝热在那里经营着一间珠宝作坊。多年之后，卡尔·法贝热接手了父亲的珠宝店。他之前在欧洲多国的丰富阅历为他之后的艺术创作提供了深厚的积淀，到1872年子承父业时已经有了丰富的商业知识和非凡的艺术眼光。那时，各家珠宝店都注重经营上流社会的生意，珠宝制作也沉浸在珠宝的尺寸和重量上，束缚在个头越大、分量越重的简单逻辑和竞相摆阔的风气中。年轻的法贝热经过思考，决心独树一帜，另辟蹊径，为了区别于其他珠宝商，法贝热将珠宝制作的重点由珠宝的克拉重量转移到追求艺术的创造性和工艺水平上。他一改过去珠宝店一味堆砌名贵材料的做法，大胆使用陶瓷、玻璃、钢铁、木材、小粒珍珠等材料。法贝热最注重的是设计，他的作品体现出哥特、文艺复兴、巴洛克、新艺术等多种风格，当他的对手们还守着传统的白色、淡蓝色及粉红色等颜料不变的时候，法贝热锐意创新，将黄、紫红、橙红和各种各样的绿色等，总共有超过140种全新的颜色运用在珠宝制作中，引领了时代潮流。

善于抓住机会，是法贝热在通往成功道路上不可或缺的关键因素。在得到沙皇赏识之前，他努力争取到皇帝艺苑工作，在那里见识了皇室历代传下来的各种奇珍异宝，同时做一些修补和估价工作，这段经历使他赢得了同行的承认和赞许。1882年，法贝热获邀参加全俄展览会，为了使自己的艺术品惊艳夺目、出尘脱俗，法贝热竭尽所能，制作了一批精美的珠宝参展，果然功夫不负有心人，他得到了丰厚的回报——一枚金质奖章，与此同时，多家报纸对他进行了报道。更重要的是，沙皇亚历山大三世和他的妻子玛利亚·费奥多罗芙娜皇后也来参观了展览，并被别致的法贝热展品所吸引。1886年，法贝热得到一个珠宝匠人所能得到的最高奖赏：被封为"皇家御用珠宝师"。1885年，他接到了那个著名的订单：沙皇命令他为皇后制作一枚复活节彩蛋。

三 皇室复活节彩蛋及法贝热艺术品

1885年是俄国沙皇亚历山大三世登基20周年，在这一具有特殊意义的复活节里，亚历山大三世想给心爱的妻子——来自丹麦的皇后玛利亚·费奥多罗芙娜准备一份特别的复活节礼物，以慰藉皇后的思乡之情。亚历山大三世想起了一名年轻的珠宝设计师——法贝热，这是因为他的作品曾经吸引过玛利亚的眼光。在复活节当天早上，法贝热向亚历山大三世呈上一只外表看上去简单无奇的复活节彩蛋。但出乎众人意料的是，白色珐琅外层的蛋壳打开后，里面竟然有黄金做的鸡蛋，鸡蛋里面是一只小巧的金母鸡，金母鸡肚子里还有一顶以钻石镶成的迷你后冠和一个以红宝石做成的微型鸡蛋。一只小小的复活节彩蛋里却暗藏玄机，可以想见这份特殊的礼物给皇后带来了无比的惊喜，玛利亚对法贝热的礼物爱不释手。亚历山大三世随即下令要求法贝热以后每年设计一只复活节彩蛋给皇室，并要求每只彩蛋必须是独一无二的。精湛工艺

图 1

图 2

加上与生俱来的艺术原创素质,法贝热年复一年地胜任挑战,自此成为沙俄宫廷的御用艺术家。

1894 年,亚历山大三世猝然长逝后将庞大的帝国留给了他的儿子尼古拉二世,继位新君照搬了父亲管理国家的政令举措,同时,对艺术品的追求上,他也遵循父亲的喜好,即每年让法贝热设计一款独一无二的复活节彩蛋的传统。继位之初,尼古拉二世命令法贝热继续为母亲制作彩蛋,随后又下了第二道命令,要求法贝热为其新迎娶的皇后亚历山德拉·费奥多罗芙娜每年设计一只彩蛋,如同父亲送给母亲一样。

得益于皇室的喜爱,构思精巧、做工华丽的法贝热复活节彩蛋,标志着法贝热将珠宝艺术提升到了文艺复兴以来装饰艺术的最高水平。他熟稔机械设计和装饰艺术,成功地将精密的机械装置与华丽的装饰集于复活节彩蛋,使得彩蛋的奢华之下玄机无限。那些流行于珠宝装饰的镀金银、宝石、贝壳镶嵌、掐丝珐琅、鎏金珐琅彩、錾刻、镂雕、浮雕等工艺,经过法贝热的设计被娴熟地运用在彩蛋装饰上,使其作品精致、灵动、奢华且充满魅力。如创作于 1897 年,尼古拉二世送给母亲玛利亚的皇家鹈鹕彩蛋(图 1),由法贝热的工匠大师米哈伊尔和丹麦微型画家约翰尼斯·津格拉夫制造。皇家鹈鹕彩蛋的蛋体和支架全部由杂色黄金制作,蛋体上圈錾刻蕨叶纹、花朵纹及飘带纹,蛋体顶部圆托上立雕振翅鹈鹕正在饲喂雏鸟。蛋体可以展开,由 8 幅袖珍象牙画框组成,内置莫斯科、圣彼得堡的风景图画,背面刻有所在慈善机构的名称和主旨铭文。彩蛋形制端庄高贵,装饰自然生动。1900 年的巴黎世界博览会上,沙皇彩蛋首次公开展出,让评委大吃一惊,法贝热的盛名由此远播整个欧洲。此后,法贝热设计制作的复活节彩蛋日趋华贵,工艺也更加细腻精湛。如创作于 1903 年的彼得大帝彩蛋(图 2),是尼古拉二世送给皇后亚历山德拉·费奥多罗芙娜的复活节礼物。这是一枚洛可可风格的黄金彩蛋,蛋体分为上、中、下三部分,玫瑰形切割钻石拼写出的"1703"和"1903"字符分置蛋体上部的两侧,中部四面卷草纹开光内分别绘有彼得大帝和他自建的小木屋、尼古拉二世和冬宫的图案。彩蛋通体黄金底色，纹饰镶嵌钻石和红宝石，奢华尊贵，彰显彼得大帝时期金碧辉煌的盛世风范。法贝热一直是时髦和高贵的同义词，贵族、皇室以及富人纷纷从法贝热公司购进艺术品。法贝热艺术品被看做上层社会中地位的象征。每当沙皇和皇后出访或在俄国四处巡游时，总是随身带着装满了法贝热珠宝的箱子，以备在适当的时候送给别人做礼物。到1896年尼古拉二世继位的时候，俄国沙皇大部分礼物都出自法贝热之手。

如今法贝热的名字已经成为奢华精美珠宝的代名词，其作品为法贝热及法贝热公司的工艺大师们集体创作的成果。在法贝热的珠宝作坊里，每一件艺术品的诞生大都经过了这样的工艺流程，在接到订单之后首先要经过详尽的构思设计，完成草图和模型，然后组织各工艺工匠包括金匠、银匠、上釉工匠、珠宝工匠、玉石工匠等进行分析讨论，研究工艺实施的可行性等问题，最后由各工种进行制作加工。在法贝热装饰艺术作品中，19世纪末20世纪初流行在俄罗斯以及欧洲的复古主义、民族浪漫主义、新艺术、装饰艺术风格均有充分体现，甚至在一些经典作品中还可以窥见来自中国东方艺术元素的影响。这种敢于对各种艺术风格采纳和借鉴的理念，不仅反映了艺术大师们的创新胆识，而且体现了他们的智慧与高超的艺术造诣。

了解了法贝热精美艺术品的历史源流，让我们走进"玲珑万象——来自美国的俄罗斯皇家法贝热装饰艺术展"，接近并感受这些艺术杰作的精湛工艺和美轮美奂的珠宝装饰，进而探寻俄罗斯艺术的辉煌。本次展览设在2015年正式启用的故宫博物院午门展区东雁翅楼展厅，展厅面积1000平方米，古代建筑外观完全保持原貌，内部则是既具有宫殿建筑氛围，又拥有现代展览设施的魅力文化空间。展览分为彩蛋、装饰珠宝、陈设用品、仿生玩物等部分，系统展示法贝热珠宝卓越的设计理念与卓越工艺，为观众献上别致精美的文化体验。

Superb Craftsmanship and Amazing Exquisiteness: The Decorative Art of Fabergé

Song Haiyang

On May 14, 2011, the Palace Museum (PM) and the Virginia Museum of Fine Arts (VMFA) signed a Memorandum of Understanding which has greatly advanced collaboration between the two museums in collection, personnel exchange, and cultural sharing. On October 18, 2014, the exhibition entitled *"The Forbidden City: Imperial Treasures from the Palace Museum, Beijing"* was opened at the VMFA. 119 pieces (sets) of imperial treasures dating from the Ming and Qing Dynasties, from the collection of the PM, offered the American audience an in-depth view of the splendor of Chinese imperial art. In April 2016, the PM will be displaying 234 pieces(sets) of work of art by Fabergé from the collection of the VMFA in an exhibition called *" Fabergé Revealed "*. This will give the Chinese audience an opportunity to view a selection of exquisite pieces from an American museum in the Palace Museum.

Peter Karl Fabergé (May 30, 1846-1920) was a famous Russian jewelry designer in the late 19th century and early 20th century. In his youth, he traveled to Germany, Italy, France, Britain and other European countries. In 1870, when he was in his twenties, he followed in his father's footsteps and took up the crafting of decorative jewelry. At an all-Russia exhibition held in Moscow in 1882, his works won instant fame and became popular with the royal families of many European countries. As an outstanding master, Fabergé was boldly innovative. At the workshops he opened in Moscow, Kiev and London, he designed and crafted a number of extraordinary works of art. The best-known among them were the Easter eggs, known as "Russian eggs". As perfect embodiment of the glory of the Romanov Dynasty, they were widely acclaimed by the royal families of Russia and other countries as artistic treasures.

I. The Eastern Orthodox Church and the Art of Painted Eggs

Eastern Orthodox Church, also known as the Orthodox Church or the Greek Orthodox Church, is a denomination of Christianity. It mainly adheres to the Christian tradition handed down from the Eastern Roman

Empire. Since it was introduced to Russia by the Grand Duke of Vladimir in 988 AD, it has made profound impact on Russia in terms of politics, society, culture, architecture, and handicraft.

Easter is the most important holiday in the calendar of the Russian Orthodox Church. It is in memory of the resurrection of Jesus Christ three days after his death on the cross and represents the return of spring and the revival of all living things. Easter celebrations include worship, traditional season's greetings, sumptuous feasts, and, of course, the exchange of sacred Easter eggs—the most typical objects of good luck for the holiday. Painted eggs first appeared during Easter in the 12th century AD, when they carried the meaning of "Jesus Christ's coming back to life and walking out of the stone grave" for revival and hope (Li Youyou: "The Culture of Russian Folk Toys"). The Easter egg was symbolic of the resurrection of Jesus Christ and the redemption of the sin of mankind—of both the mundane world and the paradise. Even in the pre-Christian era, eggs were considered to be symbolic of affluence. The iridescent colors on painted eggs suggest the glow and warmth of sunshine, and the state of abundance that comes with the return of vitality after the dreary winter. Russians call Easter eggs "pysanka"; the paint is made up of onions, beetroot, cotton and other daily materials. Besides common painted eggs, there are also eggs made of wood, glass, porcelain, bronze, or a mixture of valuable stone and metal.

For Russians, painted eggs can bring health, beauty, strength and affluence, and are indispensable for the celebrations of Easter. They preserve painted eggs at home like holy images and guardian angels. They also use them to greet newborn babies.

II. Fabergé, the Tsar's Royal Jeweler

Peter Karl Fabergé was born in 1846 in Saint Petersburg, where his father Gustav Fabergé ran a jewelry workshop. Many years later, Carl took over his father's jewelry shop. His rich experience in European countries served as profound preparation for his artistic creation. By the time he succeeded his father in 1872, he was already an ambitious young man with extensive knowledge about business and extraordinary taste in art. At the time, all jewelry shops were focused on doing business with the upper class, and the crafting of jewelry was marked by an obsession with size and weight—the crude logic of "bigger and heavier" and the vulgar rivalry in extravagance. The young Fabergé pondered over the situation and decided to blaze a new trail. For distinction from other jewelers, he shifted the focus of his craft from karat to creativity and workmanship. He eschewed the traditional practice among jewelers of heaping up precious materials and boldly adopted such things as porcelain, glass, iron and steel, wood, and small pearls. His foremost emphasis was design. His works represent a variety of styles—Gothic, Renaissance, Baroque, and the Art Nouveau. While his rivals were clinging to the traditional paint in white, light blue and pink, he introduced yellow, mauve, orange, and various shades of green—altogether more than 140 new hues for jewelry making—which led the fashion.

The ability to seize opportunities was crucial in Fabergé's road to success. Before he won the recognition of the Tsar, he managed to secure a job at the Royal Art Academy, where he saw all kinds of rare treasures that had been preserved by the royal family for generations while doing some work of restoration and evaluation. During this period, he won recognition and approval from his colleagues. In 1882, he was invited to participate in the all-Russia exhibition. He spared no effort to make a number of exquisite pieces of jewelry, which he hoped would achieve an effect of amazing and ethereal splendor. His pains were not taken in vain, for he was rewarded with a gold medal. His story was reported by several newspapers. More importantly, Tsar Alexander III and his wife Maria Feodorovna came to the exhibition and were attracted to Fabergé's novel pieces. In 1886, Fabergé gained the highest award for his trade—the title of "royal jeweler". In 1885, he received the famous order: the Tsar asked him to make an Easter egg for his wife.

III. Royal Easter Eggs and Fabergé's Works of Art

1885 marked the 20th anniversary of the coronation of Alexander III, and the Easter of that year carried a special significance. The Tsar wanted to prepare a unique Easter gift for his beloved Danish wife Maria Feodorovna to relieve her homesickness. He recalled the young jewelry designer called Fabergé, whose works once drew the attention of Maria. On the morning of Easter, Fabergé presented Alexander III with a seemingly ordinary egg. To everyone's surprise, when the white enamel shell was opened, an egg made of gold was revealed. Inside the egg was a delicate gold hen, in whose belly there was a miniscule queen's crown set with diamonds and a tiny egg made of ruby. Such intricacy within an Easter egg must have given the queen a wonderful surprise. Indeed, Maria was immensely fond of Fabergé's gift. Alexander III thereupon ordered Fabergé to design an Easter egg—a unique one—for the royal family each year. Thanks to his great skills and innate originality, Fabergé overcame the challenge year after year and became a royal artist at the Russian royal court.

In 1894, Alexander III died and left the vast empire to his son, Nicolas II. The new Tsar carried on his father's decrees and measures for governance. In artistic pursuit, he also followed his father's practice of asking Fabergé to design a unique Easter egg each year. At the beginning of his reign, he directed Fabergé to continue to make Easter eggs for his mother. Then he gave a new order, asking the artist to design an Easter egg for his new queen Alexandra Feodorovna each year, just as his father had done for his mother.

Thanks to the preference of the royal family, the ingenious design and magnificent craftsmanship of Fabergé's Easter eggs represented the elevation of the art of jewelry to the highest level of decorative art since the Renaissance. His intimate knowledge of mechanical design and decorative art led to a successful combination of precise machinery and splendid decoration in Easter eggs, giving them infinite mystery under their lavish appearance. The popular techniques for jewelry decoration—gold-and silver-plating,

Fig. 1

Fig. 2

gem and shell inlay, cloisonné, gilded cloisonné enamel, engraving, openwork, and relief carving—were ingeniously applied to Easter eggs for a charming effect of delicacy, vitality and luxury. For instance, the Imperial Pelican Easter Egg (Fig. 1) created in 1897 was commissioned by Nicolas II for his mother. It was crafted by Fabergé's master craftsman Mikhail and Danish miniature painter Johannes Zehngraf. The body of the egg and the stand are made of mottled gold. The egg is circularly engraved with designs of fern leaves, flowers and ribbons. On the circular pad atop the egg stands a pelican with spread wings feeding a young bird. The body of the egg can be unfolded to show eight tiny ivory frames containing pictures of the scenery of Moscow and Saint Petersburg. The back is engraved with the names of the charities and their principles. As a whole, the egg is refined and dignified in shape and natural and vivid in decoration. When the Tsar's Easter eggs were shown to the public for the first time at the World Expo in Paris in 1900, they amazed the judges and spread Fabergé's fame throughout Europe. Since then, Fabergé's Easter eggs became even more magnificent and exquisite. For instance, the Imperial Peter the Great Easter Egg (Fig.2) created in 1903 was an Easter gift presented by Nicolas II to Queen Alexandra Feodorovna. A gold egg in the Rococo style, it is divided into three parts from top to bottom. The numerals 1703 and 1903, made up of rose-cut diamonds, are on opposite sides of the upper part. The medallions in the middle part, bordered by scrolls, are painted with Peter the Great and the cabin he built himself, Nicolas II, and the Winter Palace. The overall golden ground and the inlay of diamonds and rubies for the designs suggest luxury and magnificence of the heyday of the reign of Peter the Great. Fabergé was always synonymous with fashion and class. Nobility, royalty, and wealthy individuals all bought objects from the Fabergé firm. Such works of art would be considered as a symbol of status among the elite. When the Tsar and his queen visited foreign countries or traveled across Russia, they would always carry a chest filled with Fabergé's jewelry to be given as gifts on appropriate occasions. By the time Nicolas II succeeded to the throne in 1896, most of the gifts of the Tsar were made by Fabergé.

Nowadays the name Fabergé has become synonymous with refined luxury jewelry. The works were the result of the combined efforts of Fabergé and the masters in the Fabergé Company. In his jewelry workshops, almost each work of art was created in the following procedure: reception of the order; careful and detailed design; drafting and modeling; the organization of specialized craftsmen (goldsmiths, silversmiths, glazers, jewelers, and lapidaries) for analysis and discussion on technical feasibility and other issues, and, finally, crafting and processing by the various workers. Fabergé's decorative works of art offer a full reflection of the styles that were popular in Russia and Europe at the turn of the 20th century—classicism, national romanticism, Art Nouveau, and Art Deco. Some classic works even suggest the influence of Oriental artistic elements from China. Such bold adoption of diverse styles not only reflects the masters' innovative daring, but also their wisdom and superb artistic accomplishment.

To appreciate the history of Fabergés exquisite works of art, let us come to "*Fabergé Revealed*". Here we can have a close look at the wonderful craftsmanship of the masterpieces and the amazing decorations of jewelry, which may lead us further on to the glory of Russian art. This exhibition will be held in the East Wing of the Meridian Gate of the Palace Museum, which was officially opened in 2015. The exhibition hall covers an area of 1,000 square meters. The original exterior of the historical structure is well preserved, while the interior is a charming cultural space that combines the ambience of palatial architecture and modern facilities for exhibition. The exhibition will be divided into several parts including Easter eggs, decorative jewelry, ornaments on display, and animal-shaped playthings. It will offer the audience a systematic view of the outstanding design and craftsmanship of Fabergé's jewelry, which will be a special cultural experience.

后 记

彼得·卡尔·法贝热（1846～1920年）是俄罗斯著名金匠、珠宝首饰匠人及工艺美术设计家。由他设计制作的"复活节彩蛋"不仅是俄罗斯珠宝史上，也是世界珠宝史上的惊世奇作。

2016年4月，故宫博物院与美国弗吉尼亚美术馆，在午门东雁翅楼展厅举办"玲珑万象——来自美国的俄罗斯皇家法贝热装饰艺术展"，呈现给观众的234件（套）精美的展品，是除俄罗斯外世界最大的法贝热珍品集萃，系统展示了法贝热珠宝卓越的理念与工艺，为观众献上别致精美的文化体验。为此我们特意编辑了这本图录，以期使展览资料得以长久保留。

弗吉尼亚美术馆成立于1936年，现有藏品超过33000件，藏品历史年限跨越5000年，涵盖了世界上诸多重要文明，是公认的美国顶尖综合性艺术博物馆之一。

展览的成功举办与图录的如期出版得益于故宫博物院和弗吉尼亚美术馆同仁的智慧与努力，双方精诚合作、配合默契，既体现了两馆卓越的专业素养，更表达了对文化艺术的尊重与理解。

在此，我们谨向为本次展览成功举办和展览图录如期出版所付出辛勤劳动的两馆工作人员致以衷心的感谢！

编委会
2016年4月

Postscript

Peter Karl Fabergé (1846-1920) was a noted Russian goldsmith, jeweler, and designer of arts and crafts. The Fabergé eggs he designed and made as Easter gifts were masterpieces in Russian as well as the world jewelry history.

In April 2016, the Palace Museum and the Virginia Museum of Fine Arts will host the "*Fabergé Revealed*" exhibition at the East Wing of the Meridian Gate Gallery. The 234 pieces (sets) of Fabergé treasure that will be presented to the audience, which constitute the largest Fabergé collection outside Russia, will systematically display the excellent philosophy and craftsmanship of Fabergé jewelry, offering the audience a fabulous cultural experience. We have edited this catalogue to permanently preserve the exhibition materials.

Established in 1936, the Virginia Museum of Fine Arts (VMFA), recognized as one of the top comprehensive art museums in the United States, has a collection size of more than 33,000 works that span 5,000 years and cover a number of important civilizations in the world.

The success of the exhibition and the timely publication of the catalogue is attributed to the wisdom and hard work of and the sincere cooperation between colleagues of the two museums, who have not only displayed their professional excellence, but also the respect for and understanding of culture and art.

We hereby extend our heartfelt gratitude to the staffs of the two museums that have worked tirelessly for the success of the exhibition and the timely publication of the catalogue.

Editorial Board
April, 2016

故宫博物院编辑出版委员会

主　任
单霁翔

副主任
纪天斌　王亚民

委　员
宋纪蓉　冯乃恩　娄　玮　任万平　李小城
方　遒　冯　辉　李绍毅　李永兴　梅　雪　史宁昌　宋玲平　孙　淼
王跃工　闫宏斌　余　辉　曾　君　章宏伟　张　荣　赵国英　朱鸿文

展览统筹
孙　淼　纪　炜

展览主持
宋海洋

图录主编
宋海洋

筹展编务
王　博　桑颖新　李怀玉　薄海昆

翻　译
李绍毅　袁　宏　王　蕾　王丝滢

Editorial Committee of the Palace Museum

Director

Shan Jixiang

Deputy Directors

Ji Tianbin   Wang Yamin

Members

Song Jirong   Feng Naien   Lou Wei   Ren Wanping   Li Xiaocheng
Fang Qiu   Feng Hui   Li Shaoyi   Li Yongxing   Mei Xue
Shi Ningchang   Song Lingping   Sun Miao   Wang Yuegong
Yan Hongbin   Yu Hui   Zeng Jun   Zhang Hongwei   Zhang Rong
Zhao Guoying   Zhu Hongwen

Exhibition Planners

Sun Miao   Ji Wei

Exhibition Curator

Song Haiyang

Catalogue Editor-in-Chief

Song Haiyang

Exhibition Team

Wang Bo   Sang Yingxin   Li Huaiyu   Bo Haikun

Translators

Li Shaoyi   Yuan Hong   Wang Lei   Wang Siying